OXFORD THEOLOGICAL MONOGRAPHS

THE PRINCIPLE OF RESERVE IN THE WRITINGS OF JOHN HENRY NEWMAN
By R. C. Selby. 1975

THE COSMIC CHRIST IN ORIGEN AND TEILHARD DE CHARDIN
By J. A. Lyons. 1982

THE HIDDEN GOD
By Samuel E. Balentine. 1983

PROTESTANT REFORMERS IN ELIZABETHAN OXFORD
By C. M. Dent. 1983

REVELATORY POSITIVISM?
Barth's Earliest Theology and the Marburg School
By Simon Fisher. 1988

THE COMMUNION OF SAINTS
Radical Puritan and Separatist Ecclesiology 1570–1625
By S. Brachlow. 1988

PROBABILITY AND THEISTIC EXPLANATION
By Robert Prevost. 1990

VERBAL ASPECT IN NEW TESTAMENT GREEK
By Buist M. Fanning. 1990

'WORKING THE EARTH OF THE HEART'
The Messalian Controversy in History, Texts, and Language to AD 431
By Columba Stewart, OSB. 1991

THE LORDSHIP OF CHRIST
Ernst Käsemann's Interpretation of Paul's Theology
By David V. Way. 1991

BEAUTY AND REVELATION IN THE THOUGHT OF SAINT AUGUSTINE
By Carol Harrison. 1992

THE MURATORIAN FRAGMENT AND THE DEVELOPMENT OF THE CANON
By Geoffrey Mark Hahneman. 1992

Christology and Cosmology

MODELS OF DIVINE ACTIVITY IN ORIGEN, EUSEBIUS, AND ATHANASIUS

J. REBECCA LYMAN

CLARENDON PRESS · OXFORD

This book has been printed digitally and produced in a standard specification in order to ensure its continuing availability

OXFORD
UNIVERSITY PRESS

Great Clarendon Street, Oxford OX2 6DP
Oxford University Press is a department of the University of Oxford.
It furthers the University's objective of excellence in research, scholarship, and education by publishing worldwide in
Oxford New York
Auckland Cape Town Dar es Salaam Hong Kong Karachi Kuala Lumpur Madrid Melbourne Mexico City Nairobi New Delhi Shanghai Taipei Toronto
With offices in
Argentina Austria Brazil Chile Czech Republic France Greece Guatemala Hungary Italy Japan South Korea Poland Portugal Singapore Switzerland Thailand Turkey Ukraine Vietnam

Oxford is a registered trade mark of Oxford University Press in the UK and in certain other countries

Published in the United States by Oxford University Press Inc., New York

Reprinted 2010

ISBN 978-0-19-826745-4

For
Maurice Wiles

ACKNOWLEDGEMENTS

This book is a revision of a doctoral thesis submitted to the Theology Faculty of the University of Oxford. For their support of the initial research at Oxford, I wish to thank the Principal and Fellows of St Hilda's College; the Theology Faculty for a Denyer–Johnson Studentship; and the President and Fellows of St John's College for a Junior Research Fellowship. For their support of revisions in the United States, I am grateful to the Board and President of the Church Divinity School of the Pacific, Berkeley.

Maurice Wiles provoked and supervised the thesis, as well as reading revisions for publication, and to him this book is gratefully dedicated. I am also indebted to my parents, The Revd Howard A. Lyman and Janyce V. Lyman, for early lessons in historical criticism and theological reflection; to colleagues and friends in Britain and the United States for their impatience as well as interest; and to Andrew Bridges, my husband, whose passion for Hellenism has been an essential stimulus and support.

CONTENTS

Abbreviations viii

INTRODUCTION 1

1. EARLY CHRISTIAN COSMOLOGIES 10
  Ancient Theological and Philosophical Perspectives 11
    Divine Transcendence and Causality 11
    Human Freedom and Redemption 28
    Conclusion 36

2. ORIGEN: GOODNESS AND FREEDOM 39
  Origen as Teacher 40
  Origen as Theologian 47
    Divine nature and will 47
    Human nature and will 58
    Christology 69

3. EUSEBIUS OF CAESAREA: POWER AND PROGRESS 82
  Eusebius as Apologist 83
  Eusebius as Theologian 90
    Divine nature and will 90
    Human nature and will 100
    Christology 106

4. ATHANASIUS: TRANSCENDENCE AND TRANSFORMATION 124
  Athanasius as Bishop 125
  Athanasius as Theologian 132
    Divine nature and will 132
    Human nature and will 141
    Christology 146

CONCLUSION 160

Bibliography 165

Index 182

ABBREVIATIONS

Complete references to works cited in the text or notes may be found in the Bibliography. Early Christian literature is cited in the notes by the standard Latin titles and abbreviations found in G. W. H. Lampe, *A Patristic Greek Lexicon*, pp. ix–xliii. Unless otherwise noted, all translations are my own.

ANF — *The Ante-Nicene Fathers*, ed. A. Roberts and J. Donaldson, American Edition (Buffalo, 1885–96; repr. Grand Rapids, 1951–6)

ANRW — *Aufstieg und Niedergang der römischen Welt*, ed. H. Temporini and W. Haase (Berlin, 1971–)

CH — *Church History*

CHS — The Center for Hermeneutical Studies in Hellenistic and Modern Culture, The Graduate Theologial Union and the University of California, Berkeley, California

GCS — *Die griechischen christlichen Schriftstellen* (Berlin, 1897–)

HTR — Harvard Theological Review

JTS — *Journal of Theological Studies*

NF — *A Select Library of Nicene and Post-Nicene Fathers of the Christian Church*, second series, tr. and ed. P. Schaff and H. Wace (London and New York, 1887–94; repr. Grand Rapids, 1952–6, 1978)

PG — *Patrologia Graeca*, ed. J.-P. Migne (Paris, 1857–86)

SC — *Sources chrétiennes*, ed. H. de Lubac, J. Daniélou, *et al.* (Paris, 1942–)

SP — *Studia Patristica*

TS — *Theological Studies*

VC — *Viligiae Christianae*

ZNW — *Zeitschrift für die neutestamentliche Wissenschaft*

INTRODUCTION

> Query: Whether Christ sent his apostles to preach metaphysics to the unlearned common people and to their wives and children.
>
> Isaac Newton, 'Queries regarding the word *Homoousios*'

The relation of Christian belief to Greek thought has been a source of controversy since the first century. Following such disparate critics as Porphyry and A. Harnack, scholars continue to debate the significance of the religious sea change of Late Antiquity.[1] Even if one puts aside the confessional or apologetic assumptions of later scholarship, the early Christians themselves offer no uniform vision of their relation to Graeco-Roman society, describing themselves as both citizens and strangers. The historical issue is the recovery of the construction of a religious world-view; but theological and historical studies alike are still plagued by the acerbic rivalry between Christians and 'pagans' preserved in the ancient texts and in the traditional interpretations of these texts. As S. R. F. Price commented in a recent study of the imperial cult in Asia Minor, 'There is a deep-rooted ethnocentric desire to play off Greek and Roman cults against Christianity so as to define its standing.'[2]

Historical revisions of the identifying oppositions seem threatening to both Christian and post-Christian assumptions about the 'rise' of Christianity and the 'decline' of classical antiquity. Some scholars therefore emphasize distinctions—for example, H. Dörrie denies any common ground between Platonism and Christianity—while others defend synthesis—for example,

[1] Helpful review articles of recent literature on this topic include H. Dörrie, 'Was ist "spätantiker Platonismus"?', *Theologische Rundschau*, NS 36 (1971), 285–302; E. P. Meijering, 'Zehn Jahre Forschung zur Thema Platonismus und Kirchenväter', ibid. 303–20; idem, 'Wie platonisierten Christen?', *VC* 28 (1974), 15–28; C. J. de Vogel, 'Platonism and Christianity', *VC* 39 (1985), 1–62; A. M. Ritter, 'Platonismus und Christentum in der Spätantike', *Theologische Rundschau*, NS 49 (1984), 31–56.

[2] S. R. F. Price, *Rituals and Power* (Cambridge, 1984), 14. Jonathan Z. Smith, *Drudgery Divine* (Chicago, 1990), also discusses motivations for the stark comparisons of Christianity and Graeco-Roman religion.

F. Ricken appeals to a scriptural base or an orthodoxy implicit in Hellenized formulas.[3] Ironically, critical historical accounts and apologetic theological studies may be marred by similar assumptions if early Christianity is defined by the teaching of later orthodoxy. The religious and theological intentions of ancient authors, particularly controversial ones, have often been obscured by evaluations based on the expectations of later creeds, rather than on the coherence of their theology in its own right; doctrinal deviations have often been categorized as intellectual, since philosophy and heresy constitute an ancient but lasting association.[4] Yet, to use categories of Scripture alone to define ancient Christianity is also not helpful to the historian. The simple, but common, contrast between biblical and Greek thought often ignores patterns of exegesis grounded in the issues of contemporary Hellenism, as indeed was the very language of the Christian Scriptures.[5] The underlying effect of religious experience or liturgy is also difficult to recover and to use critically, without presupposing orthodox sensibilities. If a particular model of devotion is assumed as normative, the possibility that a 'heresy' may have religious motivations is lost, together with the historical plurality of ancient Christian beliefs and practices.[6] The

[3] Dörrie, 'Was ist "spätantiker Platonismus"?'; F. Ricken, 'Nikaia als Krisis des altchristlichen Platonismus', *Theologie und Philosophie*, 44 (1969), 321–41; idem, 'Zur Rezeption der platonischen Ontologie bei Eusebios von Kaisareia, Areios und Athanasios', ibid. 53 (1978), 321–52. Compare de Vogel's defence of the natural affinity between Platonism and Christianity in 'Platonism and Christianity' with W. Pannenberg, 'The Appropriation of the Philosophical Concept of God as a Dogmatic Problem of Early Christian Theology', in idem, *Basic Questions in Theology*, vol. 2, tr. J. Baker (London, 1971), 119–83.

[4] On the polemical creation of the association of heresy and philosophy, see A. Le Boulluec, in *La Notion d'hérésie dans la littérature grecque IIe–IIIe siècles* (2 vols., Paris, 1985). This association continues in many modern studies: e.g. L. Barnard, *Justin Martyr* (Cambridge, 1967), 99; J. W. Trigg, *Origen* (Atlanta, 1983), 98f. Equally important, some scholars continue to contrast the speculation of 'heresy' to an assumed clarity and stability of orthodoxy. Meijering criticized Dörrie for using creeds to define early Christianity over against Platonism ('Wie platonisierten Christen', 16); de Vogel defined the content of Christianity as 'not a matter of discussion, but of faith' ('Platonism and Christianity', 28). R. Lane Fox also assumes an orthodox definition of Christianity in his historical study *Pagans and Christians* (New York, 1986): e.g. little attention to Gnosticism and heresy as 'perverse', 332, 493; Arius is described only as 'heretical', 602.

[5] See J. Barr, 'Athens or Jerusalem?', in idem, *Old and New in Interpretation* (London, 1966), 34–64; M. Hengel, *Judaism and Hellenism*, tr. J. Bowden (Philadelphia, 1974; repr. 1981), 107f.

[6] R. Williams assumed an orthodox model of piety and liturgy in his recent

fundamentals of Greek thought may be easier to outline, therefore, than the shape of the developing Christian theologies.

In their attempts to reconstruct early Christian thought on its own terms, recent historians have emphasized the assumptions shared by Graeco-Romans and Christians in Late Antiquity, yet underlined the distinctiveness of the approaches to reality found in Greek and Christian religion. Following anthropological models about the relation of religion to culture, historians such as W. Meeks and P. Brown have outlined in detail the 'overlap of worlds', as well as the differences between Christianity and traditional Hellenistic culture.[7] To understand early theological reflection as not merely parasitic on philosophy or as embryonic orthodoxy, although these are important, if second-order, considerations, we need, therefore, to investigate the particular religious perspective which gave common language a new context, and hence a new meaning. In general, ancient Christians affirmed the creating and redeeming power of one God, the universal and historical soteriological role of Jesus, and the responsibility of all individuals to repent and receive deliverance from the coming judgement. Within this admittedly broad Christian 'world-view', one may explore theological variations of this common pattern of beliefs which set early Christians apart from Judaism and Graeco-Roman religion without presupposing certain theological constructions.[8] This allows us to take seriously the legitimate plurality of early Christian reflection and devotion. As revealed by the many controversies and documents of early Christianity, various communities and individuals

study *Arius* (London, 1987), 110. For other models, see M. F. Wiles, 'Eunomius', in R. Williams (ed.), *The Making of Orthodoxy* (Cambridge, 1989), 157–69; T. A. Kopecek, 'Neo-Arian Religion', in R. C. Gregg (ed.), *Arianism* (Philadelphia, 1985), 153–79; R. Lyman, '*Lex Orandi*', in D. Pailin and S. Coakley (eds.), *The Making and Remaking of Christian Doctrine* (Oxford, forthcoming).

[7] W. Meeks, *The Moral World of the First Christians* (Philadelphia, 1986), 16; idem, *The First Urban Christians* (New Haven, Conn., 1983). P. R. L. Brown, *The Making of Late Antiquity* (Cambridge, 1978); idem, 'Late Antiquity', in P. Veyne (ed.), *A History of Private Life*, vol. 1 (Cambridge, 1987), 235–311; H. G. Kee, *Miracle in the Early Christian World* (New Haven, Conn., 1983).

[8] For a discussion of 'world-view' see C. Geertz, 'Religion as a Cultural System', in M. Banton (ed.), *Anthropological Approaches to the Study of Religion* (London, 1966), 3; idem, 'Ethos, World View and the Analysis of Sacred Symbols', in idem, *The Interpretation of Cultures* (New York, 1973), 126–41.

emphasized different aspects of the common world-view, and this resulted in diverse theologies.[9]

To recover the inner coherence and meaning of these early reflections on their own terms, it is helpful, therefore, to approach them as theological models which express a particular understanding of the common Christian world-view. As used by modern commentators, the word 'model' does not refer to a formal or systematic theology, but rather to an underlying pattern of relations between concepts. 'Models fund concepts even as concepts order models,' in the words of Sallie McFague.[10] The emphasis is on the use and relation of concepts, in order to recover their meaning to a specific thinker. By seeking to understand the distinctive use of common language through the working of an entire model, one may therefore clarify the originality and intention of the thinker, and perhaps the interests of the community.[11] In ancient Christianity, concepts borrowed from scriptural or philosophical sources were often modified by their use and relation in a particular theological model. For example, Valentinus worked within a Christian world-view, but his understanding of incarnation and creation included Gnostic assumptions about the relation of spirit and matter; yet these assumptions were changed by their inclusion in a Christian for-

[9] On the importance of the community as the context for meaning, see Mary Douglas, *How Institutions Think* (Syracuse, NY, 1986), 55f. and 91f.; cf. Meeks, *First Urban Christians*, 94f., 190–2. This is an important aspect of historical reconstruction, since controversial thinkers may be polemically typed as 'individual' over against the 'community' in teaching. On the plurality of ancient communities and individuals and the diverse use of theology within them, see F. Wisse, 'The Use of Early Christian Literature as Evidence for Inner Diversity and Conflict', in C. W. Hedrick and R. Hodgson, jun. (eds.), *Nag Hammadi, Gnosticism, and Early Christianity* (Peabody, Mass., 1986), 177–90, and R. Williams, 'Does it Make Sense to Speak of a Pre-Nicene Orthodoxy?' in idem (ed.), *Making of Orthodoxy*, 1–23.

[10] Sallie McFague, *Metaphorical Theology* (Philadelphia, 1982), 212 n. 4; cf. discussion of literature on theological models, ibid. 105f., and R. P. Scharlemann, 'Theological Models and their Construction', *Journal of Religion*, 53 (1973), 69–70.

[11] On form and function, see Kee, *Miracle*, 295. A comment of Michel Foucault is also helpful: 'But what we are concerned with here is not to neutralize discourse, to make it a sign of something else, and to pierce through its density in order to reach what remains silently anterior to it, but on the contrary to maintain it in its consistency, to make it emerge in its own complexity' (*The Archaeology of Knowledge and the Discourse on Language*, tr. A. M. Sheridan Smith (New York, 1972), 47).

mulation: the Gnostic saviour became historicized.[12] Thus, rather than contrasting the Hebraic, Gnostic, and Hellenic assumptions underlying theological reflection, we need to explore the coherence and tensions between these assumptions or concepts within the function and structure of the theological model itself. This would allow us to explore more carefully the particular meaning of common categories such as soteriology or incarnation which are often used without clear definition.

This is particularly important in understanding early Christian cosmology, which has often been seen as a borrowed philosophical construction or evaluated over against later orthodoxy. A more helpful course is to consider early cosmology as the theological model which reveals basic assumptions about the nature and relation of God and humanity. For Christians, not only did these assumptions frame discussion about the significance of Jesus; but, in turn, beliefs about Jesus shaped the resulting definitions of divine and human nature. Like other thinkers of Late Antiquity, Christians revised traditional cosmological forms to address contemporary problems regarding divine action and human life. Thus cosmology was not static; its very structure reflected theological creativity and deep religious concerns.[13] The sources of cosmological language are obviously significant; yet too much attention to origins alone may obscure their particular theological use. Thus biblical and Hellenic cosmologies have often been contrasted in terms of the categories of will and nature; yet the language of early Christians usually fell somewhere between them. For example, the concept of *creatio ex nihilo* was not clearly formulated until the late second century, so its use in evaluating the Christian identity of earlier theologians is anachronistic at best; but it is often so used.[14] In Christology, on the other hand, language about the will may well be 'biblical';

[12] Kurt Rudolph, *Gnosis*, tr. R. M. Wilson (New York, 1983), 149–51.

[13] On ancient cosmology, see H. Dörrie, 'Divers aspects de la cosmologie de 70 av. J.-C. à 20 ap. J.-C.', *Revue de théologie et de philosophie*, 3. 22 (1972), 400–5; D. Ulansey, *The Origins of the Mithraic Mysteries* (Oxford, 1989). For an overview of the relation of theology and cosmology, ancient and modern, see D. Tracy and N. Lash (eds.), *Cosmology and Theology* (New York, 1983).

[14] e.g. 'This view of Justin's is hardly compatible with the text of the Bible' (A. Dihle, *The Theory of the Will in Classical Antiquity* (Berkeley, Calif., 1982), 6). On the development of *creatio ex nihilo* as a later, non-biblical doctrine, see D. Winston, 'The Book of Wisdom's Theory of Cosmogony', *History of Religions*, (1971), 185–202.

yet it is highly suspect in relation to Christ, as conveying adoptionism—that is, a 'merely moral' rather than an essential relation between the Father and the Son in light of the later definitions of Nicaea or Chalcedon.[15] Finally, if discussions of human free will were based on Scripture, they were embedded in the rationalist vocabulary of Greek philosophy, so that no truly Christian sense of volition was possible between Paul and Augustine.[16]

From the watershed of Nicene credal orthodoxy, much of early Logos Christology has often appeared as an important, but problematic, example of Hellenization. According to some scholars, the affirmation of the identity of substance between the Father and the Son achieved in the fourth century meant a shift from philosophical concerns to primarily religious ones; it transformed the hierarchical cosmology of earlier Christian Platonism into a truly Christian ontology, because the theological axis was no longer 'cosmological mediation', but 'soteriology'.[17] From this perspective the earlier Logos theology was a false intellectual step which confused the function and person of Christ and led inevitably to the Arian crisis. Or, more positively, Logos Christology was an imperfect affirmation of the Son's soteriological or economic role, orthodox in intention, but marred in structure by a borrowed hierarchical cosmology.[18] While these

[15] For unity of will as a primitive or economic form of substantial unity, see W. Marcus, *Der Subordinationismus als historiologisches Phänomen* (Munich, 1963), 154–5; G. Aeby, *Les Missions divines de Saint Justin à Origène* (Fribourg, 1958), 184–6. By contrast, see J. M. Dewart, '"Moral Union" in Christology before Nestorius', *Laval théologique philosophique*, 32 (1976), 283–99, or R. Norris's classic study of Theodore of Mopsuestia, *Manhood and Christ* (Oxford, 1963).

[16] On the philosophical synergism of early Christian anthropology between Paul and Augustine, see, most recently, Dihle, *Theory of Will*, 112–13; G. C. Stead, 'The Freedom of the Will and the Arian Controversy', in H. D. Blume and F. Mann (eds.), *Platonismus und Christentum* (1983), 245–57.

[17] Logos is given a predominantly cosmological interpretation in A. Heron, 'Logos, Image, Son', in R. W. A. McKinney (ed.), *Creation, Christ and Culture* (Edinburgh, 1976), 61–2; Ricken, 'Nikaia', 340; idem, 'Rezeption', 349–52; A. Grillmeier, '"*Christus licet nobis invitis deus*"', in A. M. Ritter (ed.), *Kerygma und Logos* (Göttingen, 1979), 254–5.

[18] This is the common assessment of historical and theological handbooks: e.g. A. Grillmeier, *Christ in the Christian Tradition*, vol. 1, tr. J. Bowden (London, 1975), 107–13; J. N. D. Kelly, *Early Christian Doctrines* (5th edn., Oxford, 1980), 95–101; A. H. Armstrong and R. A. Markus, *Christian Faith and Greek Philosophy* (New York, 1960), 21. W. Marcus defended economic subordination

evaluations recognize that the strength of the active and mediating character of the Logos depended on a secondary status not to be tolerated in later tradition, they fail to give it much positive religious content. Both evaluations assume that Christian and philosophical concerns were opposed, or, at best, that theological reflection was inevitably parasitic upon philosophy, particularly in regard to explanations judged inadequate by later standards of orthodoxy. Thus, Logos Christology may better be understood as a solution to a philosophical problem of cosmic mediation than as an expression of Christian soteriology. This analysis obviously presupposes a certain model of divine activity or incarnation as essential to Christianity; yet, as noted by A. H. Armstrong, the problematic ideas of transcendence and hierarchy owe as much to scriptural language as to contemporary philosophy.[19]

Recent studies of the Arian controversy in the fourth century have attempted to recast the relation between cosmology and soteriology. R. Gregg and D. Groh have proposed a new view of Arianism, based on an examination of the relation between divine freedom, human self-determination, and the Arian Christ; they contrast the theology of Athanasius with that of Arius, as distinct cosmologies resting on will or substance.[20] R. Lorenz has tried to link Arian cosmological and soteriological concerns by arguing that the shape of Arius's teaching about the created Logos was an adaptation of Origen's account of the Incarnation.[21] Both these interpretations remain controversial, in part because they challenge the continuing temptation to categorize early thought along lines of later orthodoxy, including traditional oppositions between moral and metaphysical unity, or inspiration and deification, as well as a tendency to separate Trinitarian and Christological issues. For example, J. Lienhard recently attempted to reframe the fourth century in terms of two competing theological systems based on will and substance. However, the description of the losers (Arius, Nestorius, and

as 'orthodox'—i.e. not denying the essential unity of the Father and the Son—in *Subordinationismus*, 10.

[19] A. H. Armstrong, 'The Self-Definition of Christianity in Relation to Later Platonism', in E. P. Sanders (ed.), *Jewish and Christian Self-Definition*, vol. 1 (London, 1980), 95.

[20] R. Gregg and D. Groh, *Early Arianism* (Philadelphia, 1981).

[21] R. Lorenz, *Arius judaizans?* (Göttingen, 1979); idem, 'Die Christusseele im Arianischen Streit', *Zeitschrift für Kirchengeschichte*, 94 (1983), 1–51.

Pelagius) suggests their dependence on philosophy, with little pious motivation; emphasis on human initiative led to no true deification and mere moralism. By contrast, the winners (Athanasius, Cyril, and Augustine) took seriously the rule of faith, and in defence of incarnation preserved collective transformation.[22] In his recent illuminating and sympathetic account of Arius, R. Williams focused on cosmological issues, and assumed an orthodox religious scheme which made an Arian religious motivation ultimately obscure; if a mysterious divine will was unable to deify human nature, why was it so passionately defended?[23] In the work of both Leinhard and Williams a particular definition of soteriology and the orthodox view of incarnation continue to control the historical reconstruction and assessment of ancient theologies. Equally important, the notion of will, if biblical in origin, remains theologically suspect, and can only be explained in 'heretical' systems as ultimately 'philosophical'.

The purpose of the present study is to reconstruct the religious significance of various ancient models of divine activity and soteriology by re-examining the transitions of Logos Christology in the third and fourth century. Because of their common Alexandrian theological tradition, creative cosmologies, and distinct significance within Christian orthodoxy, Origen, Eusebius of Caesarea, and Athanasius have been taken as subjects. Their reflections on cosmology and Christology have often been related to contemporary philosophical categories, and have commonly been contrasted with one another as theocentric, cosmological, or soteriological. These evaluations have rested largely on perceptions regarding the source and purpose of their language about Christ. Eusebius, for example, is 'cosmological' and 'moralistic' because of his philosophical language about the Logos, by contrast with the 'biblical' and 'soteriological' interests of Athanasius.[24] However, by focusing on the relation and use of the language of will and substance in the theological models of

[22] J. Lienhard, 'The "Arian" Controversy', *TS* 48 (1987), 415–37.

[23] Williams, *Arius*, 228–9, 238–42.

[24] Frances Young, *From Nicaea to Chalcedon* (Philadelphia, 1983), 69; Kelly, *Early Christian Doctrines*, 225, 243; Ricken, 'Nikaia', 339–40. E. Mühlenberg contrasted the Christocentrism of Athanasius with Origen's and Eusebius's theological interests in 'Vérité et bonté de Dieu', in C. Kannengiesser (ed.), *Politique et théologie chez Athanase d'Alexandrie* (Paris, 1974), 215–30.

each as background to Christological definitions, I will try to cast their familiar discussions of creation and salvation in a new light. In spite of common inherited theological formulas and firm acceptance of the rule of faith, many devotional and theological aspects of ancient Christianity were not as clear or as uniform as is often assumed. By exploring the theological and devotional function of the will of the Son in these Logos Christologies, it should therefore be possible to broaden our perceptions of the sources and purposes of ancient reflection and exegesis. The languages of will and substance were not necessarily opposed as moral versus metaphysical or biblical versus philosophical, but were often blended to address problems of fatalism, incarnation, and scriptural exegesis.[25]

Equally important, these men shared a common theological heritage. But, as urban teacher, civic apologist, and ascetic bishop, they approached the construction and purpose of Christology in ways appropriate to their different roles and communities. Although I will not focus on a sociological analysis of each theology, I will use the particular social and ecclesiastical context of each author to help illuminate the force, challenges, and intention of his work. These authors also influenced one another; and from their common patterns, as well as their diversity, it is possible to gain a richer sense of ancient Christian religious beliefs and theological creativity. This examination of three Alexandrian theological models may thereby help to sharpen our understanding of ancient Christian identity and thought in light of both answers and unresolved questions regarding the nature and shape of salvation.

[25] 'Voluntarist' and 'essentialist' cosmologies were placed in opposition by Gregg and Groh in *Early Arianism*. Lienhard also contrasted two systems focused on moralism or divinization in '"Arian" Controversy'.

I

EARLY CHRISTIAN COSMOLOGIES

> For first any one might be at a loss to know who has come forward to write, whether Greek or barbarian or if there is something in between. . . . For they would see that we agree neither with the opinions of the Greeks nor with the practices of the barbarians.
>
> (Eusebius, *The Preparation of the Gospel*)

Christian reflection began in what has been described as an era of anxiety and change in intellectual life. Beneath the traditional rhythm of philosophical reflection and civic ritual, popular philosophy flourished in Hermeticism and Gnosticism, as well as in forms of Christianity and Judaism. According to one's spiritual stability, it was an age of freedom as much as anxiety.[1] In the second century, the contests between teachers of Gnosticism, Christianity, and philosophy surfaced to a literary level; and by the third century, Christians had a secure footing in society. In the wry words of R. Wilken, the efforts of the apologists succeeded in enabling Christianity to be labelled a third-rate philosophy rather than a first-rate superstition.[2]

The motivations and causes of theological reflection in early Christianity were obviously complex, but the self-definition of the community as a missionary group and the centrality of teaching

[1] On the religious temper of Late Antiquity, contrast E. R. Dodds, *Pagan and Christian in an Age of Anxiety* (Cambridge, 1965) with R. Gordon, 'Fear of Freedom?', *Didaskalos*, 4 (1972), 48–60; R. C. Smith and J. Lounibos (eds.), *Pagan and Christian Anxiety* (Lanham, Md., 1984), and most recently Fox, *Pagans and Christians*, 261: '[Christians] were joining the most extreme option in a period when religious issues were very lively.'

[2] R. Wilken, 'Wisdom and Philosophy in Early Christianity', in Wilken (ed.), *Aspects of Wisdom in Judaism and Early Christianity* (Notre Dame, Ind., 1975), 160.

as shared reflection on text and life produced the main energy for theology. In defence of their beliefs over against those of traditional religions or other Christians, various communities defined themselves as much by teaching as by ritual or structures of common life.[3] Much of the literary evidence is preserved, therefore, in forms which echo the contemporary debates of the philosophical schools; but the force of the arguments and the resulting theological models bear a closer resemblance to the faith of the community than to philosophy, as their ancient critics pointed out. Commonplace language about transcendence, fate, or morality acquired a different resonance when placed within a Christian, rather than a Platonic or Stoic, world-view. Equally important, the varieties of early Christian identity and theology were formed by conflict and dialogue with a culture whose own religious, social, and philosophical forms were in transition. In order to understand the construction of early Christian cosmologies, it is necessary, therefore, to clarify not only the changing religious and philosophical assumptions about God and humanity in Late Antiquity, but also the way in which beliefs in salvation through Jesus shaped reflection on these common concerns in the second century. As the philosopher Celsus asked, 'If you say that every person has become a son of God by divine providence, what is the difference between you and anyone else?'[4]

ANCIENT THEOLOGICAL AND PHILOSOPHICAL PERSPECTIVES

Divine transcendence and causality

Throughout Scripture, pastoral and polemical writings, acts of martyrs, and apologetic works, Jews and Christians affirmed the

[3] Meeks, *First Urban Christians*, 92–4. On the varied relations of theological teachers to their communities, see J. Lebreton, 'Le Désaccord de la foi populaire et de la théologie savante dans l'Église chrétienne du IIIe siècle', *Revue d'histoire ecclesiastique*, 19 (1923), 481–506; 20 (1924), 5–37; H. J. Carpenter, 'Popular Christianity and the Theologians in the Early Centuries', *JTS*, NS 14 (1963), 294–310; L. W. Countryman, 'The Intellectual Role of the Early Catholic Episcopate', *CH* 48 (1979), 265. F. Wisse argues that only in certain communities was consistent—i.e. orthodox—teaching valued ('Use of Early Literature'). R. Williams stresses the continual communication and historically based focus of mainstream communities in 'Pre-Nicene orthodoxy?'.

[4] Origen, *Cel.* I. 57 (*SC* 132. 230).

absolute power of one God to create the world and to accomplish a plan of salvation within it. This uncompromising confession quickly became an axiom of Christian identity and reflection: 'We believe in the sovereign Lord, maker of heaven and earth and sea and everything in them.'[5] As has often been noted, the creative and omnipotent will of God in Hebrew and Christian Scriptures was in distinct contrast to the eternal and rational causality of Hellenic cosmology; neither the philosophical vocabulary nor the conceptual categories of the latter were adequate to convey the biblical sense of a mysterious and omnipotent divine will.[6] This contrast may be misleading, however, if one then neglects the common Hellenic milieu of all reflection in Late Antiquity, including that of Jews and Christians, or the different goals of religious, theological, and philosophical language. The first Christian reflections on divine nature and action were not simply a translation of Scripture into a philosophical idiom, but rather a passionate and sometimes clumsy attempt to give ordered expression to an emerging religious world-view which was distinct from both Hebrew Scriptures and Hellenistic philosophy. Thus theological and philosophical cosmologies as models of particular world-views served different ends, even if they shared a similar vocabulary.[7]

The revival of Platonism in Late Antiquity gave a new impetus and a new shape to cosmological questions, culminating in the thought of the Neoplatonists. The philosophical schools offered a variety of interpretations of cosmic order, salvation, and communication with the divine, as the sceptical questions of Hellenistic philosophy were gathered into the scholastic and epistemological arguments between Stoics, Aristotelians, and later Platonists. The syncretistic constructions of the diverse Middle Platonists, blending Stoic and Aristotelian ideas in their attempt to present the

[5] This echo of Exod. 20: 11 and Acts 4: 24 is a standard confession in the martyr *acta*. See 'The Martyrdom of . . . Apollonius', in H. Musurillo (ed. and tr.), *The Acts of the Christian Martyrs* (Oxford, 1972), 90; 'The Martyrdom of Pionius', in ibid. 156; and, in general, see indices, 378. For pastoral and ecclesiastical applications of monotheism, see *1 Clement*, 19–21, and Ignatius, *Magn.* 3, 6–7.

[6] Armstrong and Marcus, *Christian Faith and Greek Philosophy*, 5; Dihle, *Theory of Will*, 4–19, 72–3.

[7] J. Pépin, *Théologie cosmique et théologie chrétienne* (Paris, 1964), 502–6; R. Sorabji, *Time, Creation, and the Continuum* (New York, 1983), 317; E. P. Meijering, 'God, Cosmos, History', in *God Being History* (Amsterdam, 1975), 53–79.

true meaning of Plato's *Timaeus*, provided a popular intellectual model of an eternal, hierarchical order. Stretched between the archetypal perfection of transcendent being and the chaotic flux of the material realm, this world, even as an inferior image, possessed an anthropocentric unity with an implicit optimism for individual union with the divine. Scholastic authors such as Albinus defined the first principle as an eternal mind, identified with the Platonic Good, which ensured the structure and rhythm of the entire cosmos through an active world soul.[8] In order to affirm and protect the perfection of the first principle, the notions of transcendence and the hierarchy of mediating deities were increasingly developed.[9] In Numenius, for example, the static quality and unity of the transcendent first god were sharply contrasted with the activity and multiplicity of the second, identified with the demiurge of the *Timaeus*, who was the direct cause of the material world.[10] This hierarchy of nature and causality corresponded to the levels of perfection, order, and rational action in the cosmos as a whole. Divine causality, if personified at all, was conceived of as the unhindered and rational expression of transcendent nature. The beginning of the world was an expression of the thought of the divine mind, and therefore eternal.

If philosophical reflection on cosmology in Late Antiquity was based on the tension between rationality and necessity or being and becoming, the Platonic criticism of religious cosmogony was put most vividly by Galen in the second century:

> For the latter it seems enough to say that God simply willed [βουληθῆναι] the arrangement of matter and it was presently arranged in due order; for he believes everything to be possible with God. . . . We however do not hold this; we say that certain things are impossible by nature and that God does not even attempt such things at all, but that he chooses the best out of the possibilities of becoming. (*De usu partium*, 11.14 (ed. Walzer, 12–13))

The demiurge, as the personification of the active power of transcendent goodness and reason, limited and organized the chaos and flux of eternal matter; the divine will (βούλησις, θέλησις)

[8] Albinus, *Didaskalikos*, 10. 2–3 (ed. Louis, 56–7), 14. 3 (ed. Louis, 81); cf. J. H. Loenen, 'Albinus' Metaphysics', *Mnemosyne*, 9 (1956), 296–319.

[9] A. H. Armstrong, *The Architecture of the Intelligible Universe* (Amsterdam, 1967), 11; J. Whittaker, 'Ἐπέκεινα νοῦ καὶ οὐσίας', *VC* 23 (1969), 91–104.

[10] Numenius, in *Fragments*, fr. 6, 8, 11, 16, ed. E. des Places (Paris, 1973).

was the exercise of the rational power of transcendent nature, which brought order and stability to the changeable, material world.[11] Thus, when the second-century philosopher Atticus defended the idea of creation and providence in a curiously literal interpretation of the *Timaeus*, he gave little attention to any concept of will, but rather emphasized the power (δύναμις) of the good.[12] In accordance with his emphasis on transcendence, Numenius described the first principle as beyond both will and necessity, and perhaps even thought, eternally unmoving and undivided.[13] Taking on the dynamism of Stoic cosmology, Middle Platonists affirmed divine purpose and power in the world, yet within the definitions of the transcendent and rational nature of the divine. Activity itself was thus increasingly relegated to the second principle, described as word (λόγος) or mind (νοῦς), in order to protect the transcendence and simplicity of the first principle. However, the activity of the λόγος in the world ensured accessibility to divine power and human self-determination.[14]

In the third and fourth centuries the philosophical discussion of transcendence and causality intensified among the Neoplatonists. Emanation (προβολή) was not a conscious casuality, but rather a natural overflow of self-communication. Plotinus defined the θέλησις of the One not as a desire to create, but rather a will to be without external limitation; will was simply the expression of the perfect coincidence of act and being in the transcendent One.[15] By contrast, his pupil Porphyry emphasized the outward power of the supreme god and his ability to exercise

[11] Plato, *Timaeus*, 29E, 30B, 41A–B; see L. Taran, 'The Creation Myth in Plato's *Timaeus*', in J. P. Anton and G. L. Kustas (eds.), *Essays in Ancient Greek Philosophy* (Albany, NY, 1971), 372.

[12] Atticus, fr. 13, in *Fragments*, ed. des Places. The ultimate importance of his literal interpretation remains unclear; cf. J. Dillon, *The Middle Platonists* (London, 1977), 248–50. For possible influence on Christian authors, see E. P. Meijering, '*ΗΝ ΠΟΤΕ ΟΤΕ ΟΥΚ ΗΝ Ο ΥΙΟΣ*', in *God Being History*, 81–8, and G. C. Stead, 'The Platonism of Arius', *JTS*, NS 15 (1964), 16–31.

[13] Numenius, fr. 6, in *Fragments*, ed. des Places.

[14] Dörrie, 'Divers aspects', 412.

[15] Plotinus, *Ennead*, 6. 8. 13 (ed. Henry, 291–2); A. H. Armstrong, 'Two Views of Freedom', *SP* 17.1 (1982), 397–406; J. M. Rist, *Plotinus* (Cambridge, 1967), 78–83. J. Whittaker notes the coincidence of will and being as a pre-Plotinian assertion of transcendence in 'The Historical Background of Proclus' Doctrine of the Ἀυθυποστατα', in *De Jamblique à Proclus* (Geneva, 1975), 216.

divine will in completeness.[16] Equally important, opposition to the possibility of a radical dualism led to a philosophical defence of divine will as the power to organize matter, as in Alexander of Lycopolis's work against the Manichees.[17] A concept of will as the powerful expression of rational order was thus intensified against fatalism in later Platonic cosmology, but continued to be regulated by concepts of eternity, transcendence, and order. Because of this, Platonists defended the purpose and, to a degree, the freedom of material existence by reference to the various levels of reality, moving toward the unhindered and rational action of the transcendent first principle. The growing emphasis on transcendence in Neoplatonism led to an increasing affirmation of divine power in the material world.

Although largely discredited in the second century, as materialistic by the Platonists and deterministic by the Aristotelians, Stoicism remained an important aspect of Late Antique thought on divine activity, contributing the dynamism of its *λόγος* principle to Middle Platonic cosmology and ethics.[18] Stoic cosmology rested on the relation of two eternal principles, one active and one passive. Initially labelled 'nature' (*φύσις*), the active principle by the time of Chrysippus was called 'spirit' (*πνεῦμα*), and was equivalent to divine reason (*νοῦς, λόγος*).[19] Although the theistic language of the Stoics was largely metaphorical, it testified to their attempt to structure physics as an embodiment of metaphysics; the immanence of divine power was the basis for Stoic theories of action and morality.[20] All existence was part of a single, graduated soul of being, with various qualities and characteristics based on the dynamic relationship between *πνεῦμα* and

[16] P. Hadot, 'La Metáphysique de Porphyre', in *Porphyre* (Geneva, 1965), 133–5; idem, *Porphyre et Victorinus* (Paris, 1968), 302. Porphyry's ontology had a sense of conscious divine action which Plotinus denied; cf. A. Smith, *Porphyry's Place in the Neoplatonic Tradition* (The Hague, 1974), 36.

[17] *An Alexandrian Platonist against Dualism*, ed. P. W. van der Horst and J. Mansfeld (Leiden, 1974), sec. 17. On the philosophical and theological refutations of Manichaeism in the third and fourth centuries, see S. Lieu, *Manichaeism in the Later Roman Empire and Medieval China* (Manchester, 1985).

[18] On the relation between Stoicism and later Platonism, see Dillon, *Middle Platonists*, 43–9. On later Stoicism, see A. Graesar, *Plotinus and the Stoics* (Leiden, 1972); R. Todd, *Alexander of Aphrodisias on Stoic Physics* (Leiden, 1976); J. M. Rist (ed.), *The Stoics* (Berkeley, Calif., 1978).

[19] *Stoicorum Veterum Fragmenta*, ed. J. von Arnium, 2. 634, 1027, 1091, cited in M. Lapidge, 'Stoic Cosmology', in *Stoics*, ed. Rist, 170.

[20] R. Todd, 'Monism and Immanence', in *Stoics*, ed. Rist, 159.

inactive matter.[21] This sophisticated system of mixture and tension within a monistic physics formed the optimistic background for the moralists of the Roman era. Epictetus and Seneca thus assigned to the rational purpose of the universe a religious sense, describing the natural order as the 'will' of God.[22] Divine power was expected and commonplace in accordance with rationality and goodness. In spite of their criticisms of Stoic determinism and materiality, later Platonists adapted the cosmic dynamism of the λόγος as an essential part of their own understanding of the unity and motion of the universe.

By contrast with the scholastic explanations of divine transcendence or purpose in philosophical writing, ancient religious texts emphasized the dependence of the individual or society upon divine power. Various religious practices flourished in Late Antiquity, from the ordered public celebrations of civic piety to the individual, secret initiations of Isis. While scholars no longer look for a failure of 'nerve' in the religious temper of Late Antiquity, it is true that certain tendencies such as individualism and dramatic events involving divine intervention against evil seem to run against the grain of the ordered cosmos of traditional Graeco-Roman thought.[23] Traditionally, philosophy and religion were seen as a continuum in ancient society, and the vocation of philosopher included traditional piety as a means of reaching the unchanging truth which underlay the practices of everyday religion.[24] Thus Plutarch read the myths of Isis to decipher certain cosmological principles, whereas Apuleius celebrated his deliverance from fate and magic by submission to her divine power.[25] Both would identify the will of God as an expression of goodness in accordance with the order of the universe; yet reli-

[21] cf. R. Todd, 'Monism and Immanence', in *Stoics*, ed. Rist, 150–2.

[22] *Ep*. 41. 4; 31. 5–6. On Epictetus and Seneca, see A.-J. Voelke, *L'Idée de volonté dans le stoïcisme* (Paris, 1973), 131–90; J. M. Rist, *Stoic Philosophy* (Cambridge, 1969, repr. 1980), 224f.

[23] P. R. L. Brown, *The World of Late Antiquity* (London, 1971), 51–7; Fox, *Pagans and Christians*, 326–30.

[24] P. R. L. Brown, *The Philosopher and Society in Late Antiquity*, ed. E. Hobbs and W. Wuellner (Berkeley, Calif., 1980). I. Hadot, 'The Spiritual Guide', in A. H. Armstrong (ed.), *Classical Mediterranean Spirituality* (New York, 1986), 436–59. This traditional role gave way to increasing marginalization, however; see G. Fowden, 'The Pagan Holy Man in Late Antique Society', *Journal of Hellenic Studies*, 52 (1982), 33–59.

[25] See comments of Kee in *Miracle*, 247; Ulansey, *Mithraic Mysteries*, 84–7.

gious language lent a greater personal, or even arbitrary, sense to divine power.

Increasingly in this era, the philosophical expression of transcendence and religious affirmations of divine power seemed to come together. Earlier Platonists had affirmed a way from cosmology to theology, but later philosophers such as Plotinus and Porphyry had repudiated it as mere speculation. Dörrie has described this change as a transition from 'Logos-Religion' to 'Nous-Theologie'; the activity of God in the world was no longer seen to reveal the divine essence, so connection with the divine was possible only through enlightenment.[26] As noted above, Porphyry therefore emphasized both the power of divine will and utter transcendence, to ensure continuing religious access to the divine.[27] Against contemporary fatalism, especially the Gnostics and the Manichees, philosophers continued to defend divine power and purpose in the world. However, in Ptolemaic cosmology, the formerly beneficent ordering of destiny (*μοίρα*) was now seen as the deterministic, oppressive force of fate; philosophers outlined levels of order from the cosmic providence (*πρόνοια*) glimpsed in the movements of the stars to the lower prison of fatalism.[28] Philosophical statements therefore defended divine purpose and order based on a transcendent rationality, whereas religious expressions celebrated the power of beneficence or deliverance.

Thus, the traditional opposition between biblical voluntarism and philosophical rationality contrasts not only different cultural world-views, but also the distinct approaches to reality found in religious and philosophical reflection. Biblical texts were written not as a rational explanation of the world, but rather as an expression of religious reflection and devotion. Scriptural writings did not provide an ontology, but rather portrayed a general

[26] H. Dörrie, 'Logos-Religion? oder Nous-Theologie?', in J. Mansfeld and L. M. de Rijh (eds.), *Kephalaion* (Assen, 1975), 115–36; idem, '*Formula Analogiae*', in H. J. Blumenthal and R. A. Markus (eds.), *Neo-Platonism and Early Christian Thought* (London, 1981), 45.

[27] Porphyry, *Mac. Magn.* 4. 2, cited in R. Grant, *Miracle and Natural Law in Graeco-Roman and Early Christian Thought* (Amsterdam, 1952), 131.

[28] See Dihle, *Theory of Will*, 103–6; E. A. Amand de Mendieta, *Fatalisme et liberté dans l'antiquité grecque* (Louvain, 1945). Religious aspects are discussed in J. Dillon, 'Plutarch and Second Century Platonism', in Armstrong (ed.), *Classical Mediterranean Spirituality*, 214–29.

pattern of divine transcendence and power, in descriptive rather than analytical language.[29] Jewish and Christian theologians thus had to struggle with inherited religious language in light of the developing communities and controversial issues of Late Antiquity. Philosophically sophisticated thinkers like Philo and Origen eagerly used allegorical interpretation to uncover the rational patterns hidden in religious texts; yet theological reflection even apart from the formal Hellenic schools, such as Rabbinic consideration of human action, inevitably bore some imprint of Graeco-Roman forms.[30] The content of early theological explanation was therefore distinct from that of philosophy in terms of the religious issues which it addressed, such as physical resurrection or demonic causes of evil; yet much of its analytical vocabulary and structure were inevitably removed from the text of Scripture. While this is an obvious historical point, doctrines which were a product of theological reflection on Scripture, such as that of *creatio ex nihilo*, are sometimes taken to be explicitly biblical, and are used somewhat anachronistically to evaluate early Jewish and Christian authors. In fact, early religious thinkers drew on contemporary philosophy—for example, Stoicism—to affirm the power of the divine will to alter reality as it saw fit, and, by this rational defence of sheer power, echoed the religious affirmations about creation.[31] Jewish and Christian theologians thus emphasized omnipotence to a greater degree than philosophers; but they were equally willing to present divine activity as consistent, not arbitrary, basing this view at times on the revelation of history, at times on the goodness of the divine nature. If theologians did not often echo the radical claims of mysterious divine will found in religious texts, neither did they deny its immediacy or power. Philosophical critics of ancient Judaism and Christianity such as Galen and Celsus were thus

[29] On divine nature in Scripture, see G. C. Stead, *Divine Substance* (Oxford, 1977), 166f.; C. F. D. Moule, 'The Borderlands of Ontology in the New Testament', in B. Hebblethwaite and S. Sunderland (eds.), *The Philosophical Frontiers of Christian Theology* (London, 1982), 1–11.

[30] E. Stiegmann, 'Rabbinic Anthropology', *ANRW* II. 19. 2 (1979), 487–579.

[31] Winston has argued that *creatio ex nihilo* was a product of second-century Christian reflection in reaction to Gnosticism ('Wisdom's Theory of Cosmogony'); see also idem, 'Creation *Ex Nihilo* Revisited', *Journal of Jewish Studies*, 37 (1986), 88–91. Dihle accepts it as a given, biblical doctrine, with little attention to historical development (*Idea of Will*, 167–8).

exasperated by the attempted rational defence of such extravagant claims. The theological use of philosophical formulas regarding transcendence or power was grounded in religious affirmations of a particular kind of divine activity which were often foreign to philosophers. To understand the theological significance of these borrowed and sometimes paradoxical formulas, it is necessary to examine them in the context of both the religious texts and the particular theological model.

Like their forebears the Jews, early Christians confessed one God as the creator of life and active in history; yet, in addition, they testified to the unique historical presence of this power revealed in the life and teaching of Jesus, the Christ. The various forms of New Testament Christology underline the tension between one God as the hidden source of life, goodness, light, and truth and the immediate access to God gained through Jesus.[32] Even at this point the Christian presentation of the nature and will of God is intimately bound up with the significance of Jesus. *βουλή* or *θέλημα* expressed the plan of God which had been decisively revealed in Christ.[33] Thus Jesus, by his preaching of the kingdom and his obedient life and death, gave a definite shape to the activity of God in history: 'Father, if you will [*βούλει*], remove the cup from me; nevertheless, not my will [*θέλημά μου*], but yours be done' (Luke 22: 42).[34] As in Hebrew Scripture, faith was defined as obedience to the will of God, especially as embodied in the example of Jesus. His obedience was a sign of his union with God, and perhaps the highest expression of divinity possible within the constraints of Jewish

[32] On the hidden nature of God, see 1 Tim. 6: 16; John 1: 18; 1 John 4: 12. On divine power and description, see Mark 10: 18; 1 John 1: 5; John 4: 24; John 1: 4–5. On Jesus as Wisdom, Son, Image, Agent, see Heb. 1: 2–3; Col. 1: 15–20; 1 Cor. 1: 24. For a summary of these titles in their eschatological and Jewish context, see F. Hahn, *The Titles of Jesus in Christology* (London, 1969).

[33] For *βουλή* in Luke as plan, see Acts 2: 23, 20: 27; as economy of salvation, Eph. 1: 11 and cf. Heb. 6: 17. For *θέλω* as God's desire for creation and redemption, see Matt. 20: 14; Rom. 9: 18, 22; Col. 1: 27; 1 Cor. 12: 18, 15: 38; 1 Tim. 2: 4. *θέλημα* is used most often for God's purpose for human salvation (Rev. 4: 11; Rom. 2: 18; Matt. 6: 10, 12: 50, 18: 14; Mark 3: 35), new life in receiving God's will (Matt. 12: 50; John 7: 17; 1 John 2: 17; Rom. 12: 2), and the obedience of Jesus (Matt. 26: 42; John 6: 38–40). On its rarity outside Christian literature, see J. H. Moulton, *The Vocabulary of the New Testament* (London, 1918–30), 286.

[34] See also Matt. 26: 42; John 6: 38–40.

monotheism.[35] This very authority and obedience also expressed and revealed the divine plan to his disciples. Thus, faith in the New Testament is not only obedience to divine will, but knowing and believing in the presence of the kingdom and in God's power to deliver humanity, based on the Christ event.[36] Christian appeals to divine will were therefore anchored in the testimony and story of God's act revealed in Christ as the plan of God to bring all believers through death to resurrection. So anchored, God's will was in some sense less free than in Hebrew Scripture because the historical event of the Incarnation gave a definite and decisive shape to the process of salvation; this will was revealed in history and formed the centre of the new community's life and worship.[37] As the debates on faith and knowledge from Paul to Clement reveal, this new religious testimony to the act of God in Christ was held in tension with the traditional Judaic testimony to the mystery of divine will and contemporary Greek ideals of noetic union. For early Christians the sovereignty of God in creation and religious life was matched by immediate and accessible power in Jesus the Messiah; through the Son, the Father had been revealed. Especially as expressed in cosmological Christological formulas, creation and salvation through Christ revealed a new context of existence and everyday life: 'For he has made known to us the mystery of his will [θελήματος αὐτοῦ] according to the purpose [εὐδοκίαν] which he set forth in him as a plan [οἰκονομίαν] for the fullness of time' (Eph. 1: 9–10).

This confidence in God's revealed, accessible saving will echoed

[35] John 4: 34; 5: 30. See A. E. Harvey, *Jesus and the Constraints of History* (London, 1982), 163f.

[36] Dihle emphasizes the sheer obedience of faith in the New Testament, in contrast to knowledge in the Greek tradition (*Theory of Will*, 73f.). However, he simplifies and neglects the role of revelation in the early community; e.g. the model of Christ for Christian obedience in Phil. 2: 12 (ibid. 204 n. 77). On the combination of faith and knowledge in the early Christian understanding of πίστις, see J. L. Kinneavy, *Greek Rhetorical Origins of Christian Faith* (Oxford, 1987), and D. Luhrmann, *Glaube im frühen Christem* (Gutersloh, 1976), 86–7.

[37] See comments of E. P. Meijering on the divine plan of salvation, which, to a degree, limits the freedom of the divine will, in 'God, Cosmos, History', 59f., and H. Dieter Betz on the particular shape of providence in divine fatherhood, in 'Cosmogony and Ethics in the Sermon on the Mount', in R. W. Lovin and F. E. Reynolds (eds.), *Cosmogony and Ethical Order* (Chicago, 1985), 175. Meeks discusses Pauline assemblies based on monotheism and eschatology in *First Urban Christians*, 190; cf. idem, *Moral World*, 97–123, on the development of multiple communities.

through the acts of the martyrs and early pastoral writings. On the one hand, God was transcendent, invisible, and unknowable by nature; yet, in Christ, God's nature had been revealed and received as good and merciful.[38] Confessions of divine creative power were therefore linked to the religious affirmation of God's saving power and will: 'For he called us when we were not [οὐκ ὄντας], and willed us from nothingness into being [ἠθέλησεν ἐκ μὴ ὄντος εἶναι ἡμᾶς].'[39] Apostles, martyrs, and disciples in whatever circumstances were participating in the final fulfilment of the divine plan or economy (βουλή, οἰκονομία) set in motion by the life of Christ.[40] Several points about Christian identity emerge from these early religious affirmations of divine nature and activity. First, like other texts of Late Antiquity, these texts consistently affirm the transcendence of God in order to emphasize divine power. Second, through the saving will revealed and enacted in Christ, Christians receive the divine power and love present and revealed by God throughout creation. It is not clear, therefore, that one may contrast philosophical interests in transcendence with biblical assertions of power, since the language of both was appropriate to concerns about divine activity.[41] The question is rather the shape and the purpose of the action. In the same way, to draw a sharp contrast between the Christian acceptance of revelation and grace and the Graeco-Roman reliance on the structure of the cosmos obscures the strong Christian affirmation of creation as the proper place of divine activity now revealed in Christ.[42] Christian thinkers limited and sharpened

[38] On divine transcendence, see *Diog*. 7. 2; *Ep. Poly*. 3. 2; 'Martyrdom of Saints Carpus, Papylus and Agathonice' (ed. Musurillo, 24). On God revealed as good and merciful, see *1 Clement*, 19. 2; *Diog*. 8. 8–11.

[39] *2 Clement*, 1. 8; cf. 'Martyrdom of Apollonius', (ed. Musurillo, 98); 'Martyrdom of Polycarp' (ed. Musurillo, 12); 'Martyrdom of Pionius' (ed. Musurillo, 156–8); *1 Clement*, 38. 3.

[40] Martyrdoms are in accordance with God's will 'Martyrdom of Polycarp' (ed. Musurillo, 2); 'Martyrdom of Perpetua and Felicitas' (ed. Musurillo, 112f.; 'Martyrdom of Pionius' (ed. Musurillo, 146). By divine power they are endured ('The Martyrs of Lyons' (ed. Musurillo, 68). *Re* life and death by divine will, see *1 Clement*, 23. 5, 35. 5. On the Christology of the martyrs, see E. Mühlenberg, 'The Divinity of Jesus in Early Christian Faith', *SP* 17 (1982), 136–46.

[41] Barnard contrasts power as Christian and transcendence as philosophical in *Justin Martyr*, 78.

[42] This contrast is found in Armstrong, 'Man in the Cosmos', in W. den Boer, P. G. van der Nat, and J. C. M. van Winden (eds.), *Romanitas et Christianitas* (Amsterdam and London), 7, and Dihle, *Theory of Will*, 71.

access to the divine through Jesus; yet, at the same time, they claimed that faith in Jesus ensured knowledge of God and of the true meaning of cosmological existence. Thus, neither the category of rational causality nor that of arbitrary will could do justice to the view of God developing in the various Christian communities. The latter interpreted the saving action of Jesus and continued relation to him as bringing both deliverance from evil and intimate union with God.

Against this background, the relation of divine nature and will in the Logos Christology of Justin Martyr reveals a struggle to blend Scripture, contemporary philosophy, and religious reflection in the second century. As a teacher in Rome, Justin used his philosophical education as the fundamental backdrop for his understanding and presentation of Christianity. In setting forth a reasoned case which cast Christianity as the universal philosophy, he not only proposed Christian answers to contemporary philosophical questions on the nature of God and the origin of the world; he also addressed exegetical issues with Gnostics and Jews. His audience probably consisted of converts seeking intellectual justification for their faith: 'I listened gladly to the teaching of Justin,' said Evelpistus, 'but my Christianity I received from my parents.'[43]

The philosophical influences on Justin's theology have been studied extensively. Like contemporary Middle Platonists, he taught the absolute transcendence of God through a battery of common philosophical descriptions, including apophatic theology, as well as the working of divine will in creation and redemption history through the Logos.[44] The shape of his resulting theological model rested on God's absolute goodness and direct power in relation to human nature and freedom:

> And we have been taught that in the beginning God, being good, created everything out of unformed matter [εξ αμορφου υλης] and if humans by their works show themselves worthy of his design [τω

[43] 'Acts of Justin and his Companions' (ed. Musurillo, 51).

[44] Studies include Barnard, *Justin Martyr*; J. C. M. van Winden, *An Early Christian Philosopher* (Leiden, 1971). For the use of the following theological terms, see ἄτρεπτος: *1 Apol.* 13. 4; *Dial.* 5; απαθης: *1 Apol.* 25. 2; ἄρρητος: *1 Apol.* 61. 11; *2 Apol.* 10. 8, 12. 4, 13. 4; *Dial.* 126. 2, 127. 2; *2 Apol.* 6; αγενητος: *1 Apol.* 14. 1, 25. 2, 49. 5, 53. 2; *2 Apol.* 6. 1, 12. 4, 13. 4; *Dial.* 5. 4. For θελημα as saving desire, see *Dial.* 102. 5, 116. See also C. Andresen, 'Justin und der mittlere Platonismus', *ZNW* 44 (1952–3), 157–95.

ἐκεινοῦ βουλεύματι], they are accounted worthy, and so we have received, of reigning with him, being incorruptible and passionless. For just as in the beginning he created us when we were not [ουκ ὄντας], so we think that in the same way, those who choose what is pleasing to God are by this choice accounted worthy of incorruption and fellowship. (*Apol.* 1. 10 (ed. Bayer, 33))

In this passage Justin echoes the Christian confessions of God's sovereignty and power in creation and redemption as well as the tensions between necessity and reason in the *Timaeus*. The absolute power of God in creation is linked directly to his redemptive will to enable humans by their own choice to gain union with God. But if incorruptibility and passionlessness were familiar philosophical goals, co-reigning (συμβασιλευω : 2 Tim. 2: 12) was not; and, not surprisingly, Justin underlined these concepts as 'received' or 'taught' in Christianity. Commonplace Platonic assumptions frame his discussion; yet they have been modified by their combination with the Christian affirmation of divine power. Whether or not Justin has a clear notion of *creatio ex nihilo* should not distract us from his central theological point about absolute divine sovereignty in creation and salvation. This, resting on the power of divine will, links his thinking to religious concepts and the eventual reasoning behind the doctrine of *creatio ex nihilo* far more than to contemporary Platonic cosmology. It is the continuing activity and power of divine will which will bring Christians to reign with God and bestow incorruption.[45] Justin thus assumes the contemporary hierarchy of spirit and matter, yet sees divine will as the active link bringing all creation into union with God.

As a philosopher and a Christian, Justin's account of divine immanence and power rests on the activity of the Son as Logos. His defence of divine simplicity and transcendence has sometimes led to the charge that philosophical assumptions limit divine activity; he indeed denied that God as transcendent could become visible.[46] The Father alone is αγενητος, while the Son as the first offspring is the servant of the Father's βουλη and born by his

[45] God created the world for the sake of humanity (*1 Apol.* 10, 13, 19, 28; *2 Apol.* 4; *Dial.* 5. 2, 6). Dihle criticizes Justin for teaching the divine shaping of pre-existent matter, seeing it as 'against the Biblical tradition' (*Theory of Will*, 168).

[46] Cf. *Dial.* 56, 127, and comments of Barnard, *Justin Martyr*, 78 and 83f.

θέλημα.[47] The Son is clearly divine as συνών and γεννώμενος, but different in number, though united by will (γνώμη).[48] Justin has therefore been seen by some scholars as teaching a cosmological subordination in which the Son's active role entails a secondary ontological status, as in contemporary Middle Platonism; the Platonic sense of the agent λογός has thus overridden the biblical sense of the begotten 'Son'.[49] Both explanations focus on the secondary status as problematic, and do not elucidate Justin's religious intentions in describing the form of Christ's relation to the Father as based on βουλή and θέλημα. As D. C. Trakatellis argued recently, the hierarchical pattern of action between the Father and the Son in Justin's cosmology owes as much to the language of Jesus's obedience and agency in the New Testament as to the second principle of Middle Platonism.[50] In both his cosmic and his incarnate activity the Son completed the Father's will—that is, the divine saving desire (θέλημα) and the design of the universe (βουλή); it was not by his own will, but the Father's that all were saved.[51]

Theologically, Justin created a cosmological picture which combined the providential agency of the philosophical λογος, which gave order and purpose to the universe, with religious language concerning the obedient relation between the Father and Jesus. To a degree, the use of scriptural language, particularly the descriptions of the humiliation and exaltation of Christ, strengthened the secondary status of the Son as the agent of the Father, whether incarnate or pre-existent. As the Logos submitted to the Father's will to be born, so Jesus, the incarnate Logos, was obedient to the will of God.[52] To Trypho, Justin claimed that even if one is unable to prove that Jesus existed eternally, he is undoubtedly the foretold Messiah, for he submitted to the Father's will.[53] The pattern of the Son's action, whether as cosmic Logos appearing for

[47] 1 *Apol.* 61. 1; *Dial.* 43. 1, 48. 3; *2 Apol.* 6. [48] *2 Apol.* 6. 3; *Dial.* 128. 4.

[49] For criticisms of Justin's cosmology as emphasizing 'Logos' over 'Son', see R. Norris, *God and the World in Early Christian Thought* (New York, 1965), 67; Barnard, *Justin Martyr*, 100; Heron, 'Logos, Image, Son', 50. Dörrie has noted that Platonic hierarchical degrees of divinity were in opposition to Christianity; see discussion in de Vogel, 'Platonism and Christianity', 3.

[50] D. C. Trakatellis, *The Pre-Existence of Christ in the Writings of Justin Martyr* (Missoula, Mont., 1976).

[51] *Dial.* 101–2.

[52] *1 Apol.* 63. 10, 63. 16; *2 Apol.* 6. 5–6; *Dial.* 41. 1, 67. 6.

[53] *Dial.* 48. 3. Although not as if such obedience made him 'Christ' (*Dial.* 67).

the Father to the patriarchs of the Hebrew Scriptures or as the incarnate Jesus obedient to God's command, was therefore the same.

Justin's apologetic theology thus knit together contemporary Middle Platonism with salvation history, so that the Son's cosmic agency owes as much to the biblical account of the acts of Jesus as to contemporary models of transcendence and immanence. Justin did not merely make of Jesus a universal cosmic principle; he sought to model the entire form of divine activity on the relation of the Son to the Father in Scripture, thereby producing a Christian cosmology: the transcendent Father sent the Son to redeem and transform the fallen creation. Soteriology was thus woven into the very structure of the cosmos by the nature and activity of Jesus as the Logos; the creation and teaching of the Logos, together with the suffering, death, and resurrection of the incarnate Son as the second Adam, broke the power of demonic forces. Obviously, a tension existed between the agency of the Logos and human obedience; one would not describe the action of the Son as 'free'. However, this problem was lessened in part by the emphasis on the unity of action in the Logos pre-existent and incarnate. Justin combined cosmic agency and perfect human obedience in one person who completed the divine economy as willed by God. The salvific action of suffering and death as continued submission to the Father was therefore central to the structure of the cosmology, as part of the divine economy rather than a pious belief added to Platonic ontology.[54] Justin drew heavily on philosophical categories, obviously; but the combination of divine sovereignty, the agency and incarnation of the Logos, and the deliverance of free humanity from the power of evil resulted in an essentially Christian world-view. Divine power and purpose were focused in the incarnate Christ, who revealed and embodied divine nature.

Writing against Gnostic teaching, Irenaeus of Lyons also attempted an ordered Christian reflection on divine activity. Although he criticized philosophy as a speculative human science, Irenaeus drew on common Middle Platonic ideas to affirm the transcendence and creative will of God over against Gnostic

[54] Contra Barnard, *Justin Martyr*, 125.

dualism or fatalism.[55] Like Justin, he used a series of common negative definitions to emphasize God's otherness; because of his immensity and simplicity, God was beyond human knowledge, yet by love and will God was revealed to humanity.[56] For Irenaeus the very transcendence of God guaranteed divine accessibility and power, since God was not contained by anything or separated by intermediaries from the world, but in fact contained all existence:

> If . . . he made all things freely, and by his own power, and arranged and finished them, and his will is the substance of all things [*est substantia omnium voluntas eius*], then this god is found to be the God who created all things, who alone is omnipotent and who alone is father, founding and making all things, visible and invisible, sensible and insensible, heavenly and earthly. . . . This is the Father of our Lord Jesus Christ: through his Word, who is his Son, through him he is revealed and manifested to all to whom he is revealed. (*Haer*. 2. 30. 9 (*SC* 294. 323–30))

Against the Gnostic idea of creation as an error, Irenaeus argued vehemently that physical and historical reality were the direct result of God's will. This defence of creation led to a strong emphasis on divine will, which was identical to thought and expressed without hesitation; if all powerful, it was also good, and not arbitrary.[57] Clearly, the vocabulary revealed his debt to philosophy; but, like Justin, he pushed these categories by his defence of the freedom of divine power continually active in the very structures of existence. God as God could accomplish all things: creation from nothing and the raising of the dead, as revealed in Scripture and Christian experience in Christ.[58] More

[55] E. P. Meijering, 'Irenaeus' Relation to Philosophy in the Light of his Concept of Free Will', in idem, *God Being History*, 19–30; R. Norris, 'The Transcendence and Freedom of God', in W. R. Schoedel and R. C. Wilken (eds.), *Early Christian Literature and the Classical Intellectual Tradition* (Paris, 1979), 87–100; P. Perkins, 'Ordering the Cosmos', in C. W. Hedrick and R. Hodgson (eds.), *Nag Hammadi, Gnosticism and Early Christianity*, 221–38.

[56] Cf. *Apostolic Preaching*, 6; *Haer*. 2. 13. 4, 4. 6. 3–5, 4. 20. 5: 'Homo etenim a se non videbit Deum; ille autem volens videbitur hominibus, quibus vult et quando vult et quemadmodum vult: potens est enim in omnibus Deus.'

[57] *Haer*. 4. 38. 3: 'Circa Deum autem virtus simul et sapientia et bonitas ostenditur, virtus quidem et bonitas in eo quod ea quae nondum erant voluntarie constituerit et fecerit. . . . Deo sine invidia donante quod est bonum.' Cf. ibid. 5. 4. 2, 3. 8. 3, 2. 10. 4, 2. 5. 4; Meijering, 'Irenaeus's Relation to Philosophy', 24.

[58] *Haer*. 2. 10. 2–4, 5. 4. 2, 5. 5. 2; cf. Perkins, 'Ordering the Cosmos', 234–6.

clearly than Justin, Irenaeus tied the transcendence of God to his absolute power to accomplish his will in the created world. What distinguished his theology from contemporary philosophy was not the categories, but the immediacy, shape, and power of divine will giving purpose to all levels of existence.[59]

As in Justin's cosmology, the nature and role of the Son constituted the linchpin of Irenaeus's theology of divine activity. The descriptions of the Son's activities as Logos, or Wisdom, and second Adam once again knit together religious language and contemporary cosmological structures to provide a theological model for Christian belief in Jesus. With the Spirit, the Son was one of the hands of the Father, and his generation was ineffable; he was the only begotten, the Logos and the good pleasure of God.[60] In the action of salvation, the relation of the Son to the Father is the same in pre-existence and incarnation, since he is the agent of the Father's will. Thus, the Son took flesh in order to give knowledge of the Father at the Father's command; God who was invisible by nature became visible in the Son.[61] His teaching revealed God's will for humanity, and by his perfect obedience as the second Adam, he overcame the oppression of evil: he is the one Son who completes the will of the Father.[62] As in Justin's Christology, the role of the Logos both cosmologically and in the Incarnation may be described as that of agent of the Father's will. The Son authentically reveals God's nature and intention, as well as exemplifying the true human person created in the image of God:

> In addition, we will be greater than our teacher for we suffer and bear things which our teacher neither suffered nor bore. But our Lord is the only true teacher and he is also the good and true Son of God who suffers in patience—the Word of God was made the son of man. He struggled and conquered. (*Haer*. 3. 18. 6 (*SC* 211. 362))

Christian reflection on divine nature and activity was thus an original attempt to combine philosophical explanation with religious affirmations to express the universal shape of soteriology. In the writings of second-century theologians, it was not the

[59] *Haer*. 5. 1. 3, 5. 6. 1, 2. 28. 4–5, 4. 6. 1–7; cf. Norris, 'Transcendence and Freedom', 97–9.

[60] *Haer*. 3. 9. 1, 4. 6. 6, 5. 15. 2, 5. 18. 2–3.

[61] *Haer*. 3. 17. 1, 5. 36. 3, 5. 16. 3, 5. 19. 1, 5. 21. 2.

[62] *Haer*. 3. 17. 1, 5. 16. 3, 5. 18. 2, 5. 36. 3.

vocabulary of creation or action which distinguished Christian cosmology, but the particular shape of divine activity, which focused on the direct effect of divine power and the will of God in creation and incarnation. Clearly the hierarchical divinities of Middle Platonism offered a helpful model for understanding the mediation of divine nature and activity; yet the relation of the Father to the Son in the New Testament must be taken seriously as an influence on early Christian reflection. Hierarchy *per se* cannot be cited as indicative of philosophical thought or influence. Equally important, this model of agency was essential to early religious affirmations about incarnation, as well as to the image of appropriate human response. The power of God in creation was consistently linked with the divine power to save and with human self-determination. For ancient Christians, therefore, material and spiritual reality were received as subject to God's power; yet the life and teaching of Jesus revealed the purpose and plan of the divine will as well as the shape of human response. Thus Christian reflection on creation drew on philosophical language of transcendence and eternity; yet the emphasis on redeeming will and power overshadowed the tension between spiritual and material categories. Equally important, Christ as the Logos and the active agent of the Father's will knit together the creating and saving purposes of God. In the light of the Incarnation, Christian theologians thus defended the intelligibility of the cosmic order as well as the power of the divine will.

Human freedom and redemption

The first Christian discussions about human nature were prompted by controversy, and drew on both religious insights regarding transformation and obedience and philosophical arguments regarding determinism. In a traditional contrast between the faith and obedience of biblical humanity and the intellectual nature of philosophical humanity, these early reflections have sometimes been evaluated as an unreflected synergism or as traditional classical moralism.[63] However, although theologians drew on contemporary ideas about human nature in order to discuss human action and reflection, the entire pattern of the relation of

[63] Dihle, *Theory of Will*, 107–8.

God and humanity in the developing Christian cosmology repeatedly cast philosophical language about the intellect in a voluntarist light. The dependence of created being on God the creator, together with a sense of disjunction due to the deliberate error of sin, led to the development of a particular view of human action. This view was expressed most clearly not in the biblical or philosophical categories used, but in the whole world-view of human creation and redemption taught by early Christian authors.

In Late Antiquity the question of human self-determination in what was increasingly considered to be a deterministic universe was a major religious and philosophical issue. In order to safeguard the moral responsibility of individuals against various forms of determinism, philosophers and religious thinkers often used the same arsenal of arguments.[64] Virtue within the Platonic tradition had been defined as action in accordance with the intellect, the mirror of the ideal world; knowledge of the Good allowed one to discipline the body and conform with the order of being, whereas evil was involuntary or due to ignorance.[65] In Stoic thought, debates about human action were centred in monistic physics, so that one was free only in so far as one knew the limits and possibilities of the cosmic order; the self was therefore identified with the immanent tension of the soul as *προαίρεσις*, or *voluntas*. For the later Stoics, who were predominantly concerned with moral life, *ἡγεμονικόν* became almost equivalent to *προαίρεσις*.[66] This account came under attack from Aristotelians and Platonists alike, as materialistic and deterministic; for Alexander of Aphrodisias, *προαίρεσις* rested on the innate rational freedom of the individual and the discontinuities of non-being within the natural order.[67] In Middle Platonism the goal of the philosophical life, achieved through physical and

[64] Amand, *Fatalisme et liberté*.

[65] N. Gulley, 'The Interpretation of "No one does wrong willingly" in Plato's Dialogues', *Phronesis*, 10 (1965), 82–96; Dillon, *Middle Platonists*, 43–5.

[66] On this voluntaristic knowledge in Seneca and Epictetus, see Rist, *Stoic Philosophy*, 223–4, and C. Stough, 'Stoic Determinism', in Rist (ed.), *Stoics*, 211–13. Emotions were no longer seen as irrational, but as disobedience to the logos; see A. L. Lloyd, 'Emotion and Decision in Stoic Psychology', ibid. 244. The imperial moralists are discussed at length in Voelke, *L'Idée de volonté*.

[67] The Aristotelian critique of Stoicism was developed at length by Alexander of Aphrodisias in his *De Fato*, 14–15 (ed. Fitzgerald, 66–79). See R. Pack, 'A Passage in Alexander of Aphrodisias relating to the Theory of Tragedy', *American Journal of Philology*, 58 (1937), 42–3.

moral discipline, was an unhindered intellect which offered the closest resemblance to the transcendent activity of the divine. Against the Stoics, the Middle Platonists argued that choice was not part of a causal chain, but that acts of will belong in the category of the possible, which is indeterminate. Although it required effort and the power of virtue to become divine, it was the mind, by means of contemplation, not the will, which eventually united one with the Forms.[68] They also outlined levels of fate corresponding to their hierarchical ontology; the stars moved in accordance with the rational and unhindered *πρόνοια*, whereas the lower levels of created being were subject to the oppressive and deterministic *εἱμαρμένη*.[69] The contemporary philosophical vocabulary consisted, therefore, of terms expressing intellectual intention rather than unhindered power. Virtue consisted in knowing and conforming to the order of transcendent perfection.

In the later Platonic tradition these problems of moral life were focused on the ontological tension between progress and stability. Given the shared nature of the soul and the intelligibles, the descent and ascent of the individual seems almost artificially imposed on the ideal hierarchy.[70] For Plotinus the descent of the soul was a counter-rational act, as the soul was seduced by pleasure and wished to be self-supporting; he used *αὐτεξούσιος* rather than *προαίρεσις* to describe the descent. His refusal to countenance a deliberate choice of a lesser good by the soul reflects the relation of knowledge and act in Platonic morality; error or evil is involuntary or the result of ignorance. Yet volitional action was here associated with movement and multiplicity, and was therefore negative.[71] For Plotinus the difficulty of the spiritual life lay in the opposition of the higher and the lower soul; the world was a stage for spiritual activity, and the lower stages of the ascent would be annihilated in the final union. By contrast, Porphyry, who described the descent of the soul as in accordance with divine necessity and purpose, emphasized the importance of time and historical event in a teleological process which would

[68] Albinus, *Didaskalikos*, 31. 2; J. Rist, 'Prohairesis', in *De Jamblique à Proclus* (1975), 103–17.

[69] Dillon, *Middle Platonism*, 44, 320–6.

[70] Smith, *Porphyry's Place*, 40.

[71] Plotinus, *Ennead*, 3. 2 (4) 3–7, cited in Rist, 'Prohairesis', 109. Cf. J. Dillon, 'The Descent of the Soul in Middle Platonic and Gnostic Theory', in B. Layton (ed.), *The Rediscovery of Gnosticism*, vol. 1 (Leiden, 1980), 364; Dihle, *Theory of Will*, 114–15.

not be repeated.[72] In either case the whole character and destiny of the soul were expressions of the intelligible archetype after overcoming the constraints of material existence. Philosophical criticism of Gnosticism, fatalism, and dualism was therefore sharp.[73]

By contrast, religious language tended to be at once more voluntaristic and less self-determining. On the one hand, individual response to divine power or communication was emphasized, as in the Rabbinic saying 'Everything is in the power of heaven except the fear of heaven—God can do everything except make a person religious.'[74] On the other hand, testimonies to the power of divinity affirmed a scheme of election or destiny.[75] Both sorts of statements relate more to problems of theodicy and the religious experience of salvation than to an analysis of human action. Because religious language focused on divine power, whether as underlying the stable cosmic order or as a hidden will, human action, even if defined as free, was usually described in a dependent relation, like bits drawn to a magnet. Thus, the obedience of Lucius and his dependence on the delivering power of Isis were both included in the final chapter of *The Golden Ass*.[76]

Whether in relation to divine will or to a natural order, in Late Antiquity all discussions about human nature and freedom took place in an atmosphere of 'soft determinism', although the sense of evil as an independent power in Judaism and Christianity clearly sharpened the saving power of God and the necessity for individual reliance upon it.[77] Up to a point, therefore, the sheer voluntarism sometimes cited as a definition of faith is more

[72] Smith, *Porphyry's Place*, 69–70.

[73] Plotinus, *Ennead*, 2. 9; A. H. Armstrong, 'Gnosis and Greek Philosophy', in B. Aland (ed.), *Gnosis* (Göttingen, 1978), 99.

[74] Saying of Rabbi Hanina bar Papa in *Niddah* 16b, cited in G. F. Moore, 'Fate and Free Will in the Jewish Philosophies according to Josephus', *HTR* 22 (1929), 381.

[75] On predestination in *Eccles.* and ascetic circles, see *Eccles.* 1: 14–15; Josephus, *Ant.* 13. 1–5.

[76] *Metamorph.* 11. 6: 'Vives autem beatus, vives in mea tutela gloriosus . . . quod si sedulis obsequiis et religiosis ministeriis et tenacibus castimoniis numen nostrum promerueris, scies ultra statuta fato tuo spatia vitam quoque tibi prorogare mihi tantum licere.'

[77] The phrase is D. Winston's, in 'Freedom and Determinism in Philo of Alexandria', *Studia Philonica*, 3 (1974–5), 47–70. On the evil imagination in Judaism, see Stiegmann, 'Rabbinic Anthropology', 527; Brown, 'Late Antiquity', 254f.

properly a description of religious experience, rather than of the normal pattern of religious moral life. Dihle, for example, recently contrasted Greek freedom, as consisting in conformity with the order of being, with Christian obedience to the mysterious divine will.[78] This contrast unfortunately ignored the pattern of life revealed in the incarnate Logos, who brought both knowledge and freedom through baptism to ancient Christians. A more significant contrast lies in the source of human weakness: material being in Platonism, disobedience in the Judaeo-Christian tradition.[79]

Drawing on standard philosophical arguments in defence of moral responsibility, some Christians in the second century defended individual freedom over against contemporary fatalism or Gnostic election theology. Created in the image of God, humanity was *αὐτεξούσιος*—that is, self-determined or moved—and hence essentially free. However, Christians broadened this Stoic term by their cosmological context; created being, as a product of divine will, was *τρέπτος*—that is, changeable and contingent.[80] Humans were thus less stable ontologically, than they were in philosophical anthropology, even though created in the image of God; for, if the definition of image included rationality, it was nevertheless a created rather than an eternal element. The problem in Late Antiquity, therefore, both ontologically and morally, was the relative stability of human nature as created in the image of God, yet clearly limited by finitude and inherited sin. To an extent, the difficulty was lessened in Christianity, since humans were intrinsically dependent on divine interaction in order to achieve consistent action. The activity of Christ, based on power, freedom, and grace, knit together the demands of obedience in Hebrew Scripture with the intellectual conformity of Greek philosophy.

Inserting philosophical terms of knowledge and self-determination in a Christian world-view, theologians gave rational categories a more voluntarist connotation. Thus Theophilus of Antioch followed Hellenistic Judaism in describing humanity as

[78] Dihle, *Theory of Will*, 71, 113.

[79] R. Norris, 'The Problem of Human Identity in Patristic Christological Speculation', *SP* 17.1 (1982), 147–59.

[80] W. Telfer, '*ΑΥΤΕΧΟΥΣΙΑ*' *JTS*, NS 3 (1957), 123–9; Stead, 'Freedom of Will', 246–9; Amand, *Fatalisme et liberté* 192–3.

neither mortal nor immortal, but free to choose either life or death. By nature, as created by God, humans are free.[81] The consequence of this definition of humanity as radically dependent yet possessing self-determination led to the understanding of human life and history as a process of interaction between God and creatures. Even though Adam and Eve were cast out of Paradise by disobedience, they fell into a world which God had prepared in order that they might redeem their sin; growth in obedience gradually occurred in accordance with attention and response to God's will and revelation in the world, most centrally through the activity of the Logos, or Wisdom.[82] Obedience and knowledge were thus both operative. Many Christian thinkers therefore structured their cosmology around the tensions between the divine plan to bring humanity voluntarily into union with God and the continuing alienation of free creatures from God. By preaching that alienation is brought about by evil in the form of demons and death, but that redemption is readily accessible through the revelation and resurrection of Jesus, they dramatized and focused the anthropocentrism of the religion of Late Antiquity. Because the human person was the locus of supernatural struggles in a cosmos, the whole of which was being pulled and lured towards salvation, human achievements in virtue could be valued most highly, while still being discussed in the light of the soft determinism of the salvific Logos.

The fragility and centrality of the free human person in this economy of salvation may be seen in different ways in the reflections of Justin and Irenaeus. Like Theophilus, Justin defined humanity as *αὐτεξούσιος*, and defended the moral responsibility of humans and angels.[83] In line with his emphasis on God's creative will, Justin outlined a process of choice and knowledge which delivered one from corruption.[84] Although clearly based on philosophical anthropological concepts such as *προαίρεσις*, this process was not the strenuous philosophical contemplation of the few, but rather the reception of divine power and response to divine will of the many through the example and teaching of Christ, the incarnate Logos. The process consisted in a combination of rational discernment and obedience, since it rested on the

[81] *Autol.* 2. 24f. [82] *Autol.* 2. 24–6.

[83] Amand, *Fatalisme et liberté*, 201–3; *1 Apol.* 102. 4; *2 Apol.* 7; cf. *1 Apol.* 43.

[84] *1 Apol.* 43. 1–2, 58. 2–3; *Dial.* 93. 2; cf. Dihle, *Theory of Will*, 16.

revelation of God's will in salvation history through the actions of the Logos.[85] Thus Justin described baptism as 'illumination'; but this comes about only after one has been convinced and has believed and fasted.[86] The work of Christ brought not only knowledge of good and evil, but also the power or will to have a relationship with God: 'For the one who loves God with all the heart and with all the strength being filled with a God-fearing mind, will reverence no other god . . . though you have means of understanding, yet you will not.'[87] In other words, Justin altered the borrowed philosophical conception of moral responsibility for enlightenment by heightening human dependence not merely on intellectual awakening to the immanent Logos, but also on the power of Christ which enabled knowledge and virtue in the individual. In the framework of the story of God's intentional plan of creation and redemption, the innate rational element as the image of God in humanity was pressed in a more voluntaristic direction to include repentance, contemplation, and obedience in relation to God. The result was neither a blindly obedient faith nor an intellectual illumination, but a working together of divine persuasion and human freedom revealed and achieved through participation in Christ.

A more explicit and similarly syncretistic pattern of human freedom was drawn by Irenaeus against Gnostic anthropology. According to their critics, Gnostics taught a mystical élitism: only those who possessed a certain knowledge could be saved, and these were often described as predestined by nature. Recent evaluations of Christian Gnosticism have suggested that in fact some of its adherents may have taught a form of Pauline election theology.[88] For them, divine will was a force of hidden deliverance within the fallen cosmos, giving 'rest' (ἀνάπαυσις) to the believer in the midst of the fallen world. The classes of nature may have referred to the stages of conversion, as well as election: 'If you are oppressed . . . and you do (the Father's) will, he will love you . . . and reckon you to have become beloved through his providence by your own choice.'[89] Characterizing the Gnostics as

[85] *Dial.* 8, 92, 131; *1 Apol.* 45. [86] *1 Apol.* 61, 65.

[87] *Dial.* 93.

[88] U. Bianchi, 'Anthropologie et conception du mal', *VC* 25 (1971), 197–204.

[89] *Apoc. of James*, 1. 4. 31–5. 10, in J. Robinson (ed.), *The Nag Hammadi Library in English* (3rd rev. edn., San Francisco, 1988), 32. 'And the will is what

fatalists, Irenaeus argued that all were created free by nature, and hence able to choose salvation. He went so far as to argue that this freedom, or self-determination (αὐτεξούσιος), was likeness to God, and could be developed in relation to God to overcome created weakness:

In his kindness he bestowed goodness and made people like himself in their own freedom [*suae potestatis*]; although through providence he knew the weakness of human beings and its consequences, but through his love and power he shall overcome created nature. (*Haer*. 4. 38. 40 (*SC* 100. 960))

αὐτεξούσιος thus describes the self-determination of a created person in relation to divine will.[90] Like Theophilus, Irenaeus describes the Fall as part of an early, imperfect stage of human existence; Adam and Eve as children were immature when tempted, and needed afterwards, to grow into an adult relationship with God.[91] He noted that all creatures are children of God by nature, but must also be such by obedience.[92] History was therefore the interaction between God's desire to lead free creatures towards this goal and the development of human love, knowledge, and faith in God. Jesus as the second Adam brought freedom from sin, as well as knowledge of God, so that creatures may at last be drawn into immortality and union with God.[93] Irenaeus's account tried to hold together the created instability of creatures and the continuing activity of the divine will. His analysis of human freedom therefore emphasized both the innate free potential of humans and their reliance on divine will; this relative freedom reflects both religious and philosophical constraints. The stability acquired through salvation, however, was not a static perfection or a purely rational activity, but rather an intimate relationship with God which bestowed not only knowledge, but

the Father rests in and is pleased with. Nothing happens without him nor does anything happen without the will of the Father, but his will is unsearchable. His trace is the will and no one will know him' (*The Gospel of Truth*, 1. 37. 20–34, ibid. 49; cf. *Gospel of Philip*, 53. 1–10, ibid. 142, on the voluntary sacrifice of Christ).

[90] *Haer*. 4. 37. 4, 4. 15. 2, 4. 37. 5, 4. 39. 3.

[91] R. F. Brown, 'On the Necessary Imperfection of Creation', *Scottish Journal of Theology*, 28 (1975), 17–25.

[92] *Haer*. 4. 41. 2; E. P. Meijering, 'Die "Physische Erlösung" in der Theologie des Irenaeus', in idem, *God Being History*, 39–59.

[93] *Haer*. 4. 4. 3, 4. 6. 5, 4. 11. 4, 4. 38. 4.

also the likeness and love of God. The final goal was the realization of the free self as it was created to be.[94]

Christian reflection on human action and nature in the second century thus drew on common philosophical arguments regarding freedom and determinism to safeguard both moral responsibility and the continual activity of divine will. However, such reflection in the context of both divine creation and the activity of Christ pushed Christian anthropology in a voluntarist direction, in spite of its dependence on philosophical categories. *αὐτεξούσιος* became a standard Christian anthropological term which underlined the self-moved character of human beings in relation to God, subject neither to fate nor to nature, and morally responsible for their salvation. However, on account of their contingent relationship with God, creatures were essentially unstable, and found true fulfilment and stability only in unity with the divine will. The incarnate Son thus became the hinge between divine power and human salvation as he revealed the Father's plan and nature, became the model for human virtue, and defeated the powers of evil and death, thereby allowing virtue and union with God to be achieved. By emphasizing the obedient actions of Christ on both cosmic and historical levels, Irenaeus and Justin gave a dynamic and universal sense to the struggles of the Christian life in following the will of the Father. The identification of Jesus as the divine Son meant that the will of God was not entirely a mysterious object of faith, but rather a visible, accessible plan of salvation now revealed in the life of the Christian community.

Conclusion

In response to Celsus's question at the beginning of this chapter, the difference between Christians and everyone else lay in the Christological contours of the former's religion, which, by portraying and communicating divine power and human potential, granted all persons the possibility of virtue and participation in the transcendent God. As seen in this brief overview of philosophical and religious perspectives in Late Antiquity, the origin-

[94] *Haer* 5. 8. 1–2, 5. 27. 2, 5. 29. 1. Stead contrasts the optimism and pessimism in Irenaeus's account of relative self-determination in 'Freedom of Will', 249–50.

ality and integrity of Christian responses to religious issues of transcendence and fatalism in the second century lay in a reworking of both scriptural and philosophical categories. *Contra* Dörrie and Barnard, a notion of hierarchical divinity was not foreign to early Christians and was not necessarily incorporated into early theology as a Platonic assumption; it in fact served soteriological purposes, and reflected some problems of scriptural exegesis. Of course, those Christians later described as 'modalist' or 'adoptionist' solved these problems of hierarchy and exegesis by a different model. Platonic cosmologies were obvious sources for Christian reflection; yet these models were themselves historicized and altered in being used to describe the relationship between Father and Son. The sense of agency and hierarchy was to a degree increased by the Christian language of obedience of the Son, which was drawn from Scripture and applied to the cosmic logos. Jesus as Logos, therefore, not only underscored the direct involvement of the transcendent God in the world but also provided a model of human response to God. Similarly, creation was discussed not with explicit reference to *creatio ex nihilo*, but as the affirmation of God's power and purpose to bring salvation and transformation to an unstable, contingent world.

Likewise the first reflections on Christian anthropology drew on contemporary intellectual categories; but, in relation to creation and salvation history in Christ, these were given a voluntarist interpretation which stressed the need for conversion of the individual from sin and evil towards true being in line with divine will and purpose. To present such a picture, neither the mysterious will and obedience characteristic of Hebrew Scripture nor the Hellenic rational cosmology sufficed. Christians were both ontologically changeable and free, obedient, and knowing. In response to contemporary determinism and the election theology of some Gnostics, the moral self-determination of the individual was held by some Christians to be his or her primary natural gift from the good and free God who has created and redeemed all existence. Knowledge and stability, however, were gained only through the incarnate Logos.

The philosophical categories of transcendence and rationality remained; but they were modified by their use in a model which affirmed God's continual, personal power in the structures of creation, most supremely through the life and teaching of Christ.

This central act allowed Christians to blend religious authority with elements of philosophical coherence. The natural order, if not united to the first principle by nature, was fundamentally tied to God by the intention and continual working of the divine will. If creation by divine will resulted in discontinuity with the world, a greater unity was achieved by the accessibility to God's power poured out on all people through Christ. This central, decisive event of revelation and power transposed borrowed language into a distinctively new religious key. As Armstrong has commented, confidence in revelation allowed Christians a density and force of interpretation that philosophical schools never had.[95] The Christian world-view was therefore distinguished by the shape and confidence in divine power accessible through the incarnate Logos, Jesus Christ, the Son of God.

[95] Armstrong, 'Self-Definition', 88.

2

ORIGEN: GOODNESS AND FREEDOM

Was Christ also . . . subject to fate according to the movement of the stars by his birth, and therefore did and suffered all these things?

(Origen, *Commentary on Genesis*)

Traditionally, Origen has been 'suspected of a great orthodoxy'.[1] His Christian commitment was unquestionable, but his theological conclusions stimulated passionate apologetic or repudiation; he was too right to be wrong, or too attractively wrong to be ignored. Controversial in his lifetime, Origen and his work have passed through centuries of scholarly revision and ecclesiastical judgement. At present he is no longer reducible to Middle Platonic, Stoic, or Gnostic elements; and even his ecclesiastical loyalty, if nuanced, is secure. In recent evaluations of his life and work, his intellectual passion has received the most emphasis.[2]

The importance of the Logos paradigm in Origen's Christology can hardly be overstressed. The accommodation of the incarnate Logos as the essential, intellectual, spiritual link between the uncreated Father and the various levels of created being was the centre of his thought and method. Extensive studies of the philosophical, soteriological, and exegetical aspects of his Christology continue to reveal the profound and problematic creativity of his

[1] Charles Williams, *The Descent of the Dove* (New York, 1939), 37.

[2] Helpful surveys of Origen studies are R. J. Daly, 'Origen Studies and Pierre Nautin's *Origène*', *TS* 39 (1978), 508–19; U. Berner, *Origenes* (Darmstadt, 1981); and H. Crouzel, 'The Literature on Origen 1970–1988', *TS* 49 (1988), 499–516. The most recent critical biographies are P. Nautin, *Origène* (Paris, 1977); J. W. Trigg, *Origen* (Atlanta, 1983); H. Crouzel, *Origen*, tr. A. S. Worrall (San Francisco, 1989).

thought.[3] If the Son's mediation is axiomatic, distinct subordination seems to be the necessary corollary. At the same time the centrality of the Logos seems to obscure the authenticity or significance of the human nature. However, the troublesome relationship of divinity to humanity in Origen's Christology reveals most clearly the theological and spiritual concerns of a third-century Christian leader.

ORIGEN AS TEACHER

As an ascetic intellectual in Alexandria, Origen's relationship to the Christian community lay in the independence of a *didaskalos* rather than in the ordered ecclesiastical hierarchy. As a philosopher, he sought to explore the timeless principles which underlay religious practice and teaching, and to teach these to the able few. This is not to cast him too sharply against the larger community. The creativity, diversity, and theological power of Christians in Alexandria is well known, and Origen's controversial success reflects the intellectual and cultural diversity of his community.[4] However, he deliberately tested philosophy against Scripture in order to present a convincing account of the world to educated Christians and perhaps to interested philosophers. As a theologian and spiritual teacher, he hoped to work out answers to contemporary questions within the framework of given theological propositions. His context thus embraced both the urbane cultural pluralism of the greatest city in the Mediterranean and the sacramental singularity of Christian life. The small groups of men and women who gathered with him to explore the exegetical and psychological puzzles relating to spiritual progress shared his own intellectual and religious passion. Although in Caesarea Origen as a presbyter would preach to larger congregations, his style and focus continued to rest on the education and development of individual insight.[5]

[3] Major studies include H. Koch, *Pronoia und Paideusis* (Leipzig, 1932); H. Crouzel, *Théologie de l'image de Dieu chez Origène* (Paris, 1956); M. Harl, *Origène et la fonction révélatrice du Verbe incarné* (Paris, 1958); G. Gruber, *ΖΩΗ* (Munich, 1962).

[4] For the most recent work on Christianity in Egypt, see B. A. Pearson and J. E. Goehring (eds.), *The Roots of Egyptian Christianity* (Philadelphia, 1986).

[5] On Origen as teacher, see J. W. Trigg, 'The Charismatic Intellectual', *CH* 50

From the first, Origen defended the use of philosophy in theological reflection, even if this caused tension within the Christian community. In his early works *On the Resurrection* and the *Stromateis* he argued for the necessity of a spiritual understanding of the claims of Scripture. These works were received as Platonizing, which they probably were.[6] His desire to be a Christian philosopher led him to travel and talk with other Christian teachers, in Rome in 215 and later in Athens.[7] Allegorical exegesis was the cornerstone of his philosophical theology; learned from a Jewish Christian perhaps, it unlocked the text to reveal the patterns of the intelligible world. Although Origen's bold use of Scripture and philosophy was attractive to former Gnostics like Ambrose, it was equally susceptible to misinterpretation. *On First Principles*, an outline of Christian physics, was probably written to defend his method of allegorical theology and to claim that his manner of enquiry was in direct succession to the apostolic teaching.[8]

The constant furore in Alexandria caused Origen to consider moving to Athens, but ordination by friendly bishops in Palestine led him to settle in Caesarea. His conflict with Demetrius seems to have stemmed from a combination of doctrinal disagreement and a struggle with authority; as a spiritual teacher, he felt little inclined to obedience or deference to clerical office, if he saw the person as lacking in theological insight or spiritual gifts.[9] Inevitably, this attitude led to problems in Caesarea with the

(1981), 13f.; P. R. L. Brown, *The Body and Society* (New York, 1988), 161; R. Valentasis, *Spiritual Guides of the Third Century* (Minneapolis, 1991). Crouzel underlines Origen's ecclesiastical loyalty in *Origen*, 1–36.

[6] Jerome described the *Stromateis* thus: 'decem scripsit Stromateas, Christianorum et Philosophorum inter se sententias comparans; et omnia nostrae religionis dogmata de Platone et Aristotele, Numenio, Cornutoque confirmans' (*Ep.* 70.4 (*PL* 22, col. 667)). The extant fragments confirm Origen's interest in philosophy and the necessity of allegorical exegesis; see R. Grant, 'The *Stromateis* of Origen', in J. Fontaine et C. Kannengiesser (eds.), *Epektasis* (Beauchesne, 1972), 285–92; idem, 'Theological Education at Alexandria', in Pearson and Goehring (eds.), *Egyptian Christianity*, 178–89.

[7] Nautin, *Origène*, 418; *HE* 6. 14. 10.

[8] *Princ.* praef. 3 (*SC* 252. 78–9); cf. Nautin, *Origène*, 424.

[9] Trigg, 'Charismatic Intellectual', 13f. Events and sources are summarized in Crouzel, *Origen*, 17–24. Bishops Alexander of Jerusalem and Theoctistus of Caesarea were critical of Demetrius's suggestion that Origen needed ordination in order to preach. Ironically, Origen was the first Alexandrian teacher clearly sanctioned by the bishop in his activities.

bishops who supported him. If his interests in Alexandria had been largely anti-Gnostic and foundational, in Caesarea he began a regular cycle of preaching and writing on Scripture. Again, his intellectual methods—allegorization, theological speculation, and the use of Jewish Old Testament texts—stimulated criticism.[10] He continued to travel, to arbitrate in theological disputes, to write works on prayer and apologetics useful to his patron, Ambrose, and to give intellectual and spiritual direction to young men and women. In a period of emerging Christian identity, Origen, as the brilliant son of a martyr, became the representative of an intense, sophisticated belief. Yet, even his painful death during a persecution did not lay to rest the controversy surrounding his theological reflections.[11]

As the core of his pedagogical and theological method, the profound eclecticism of Origen has been at the same time one of his most attractive and most repelling aspects. His use of Platonic wineskins for Christian theology has troubled both admirers and critics since his first works in Alexandria. Like many of his contemporaries, Christian and pagan, he was an eager, fearless borrower, and to his critics expressed a cool confidence in the efficacy of such a method. In his letter on his life in Alexandria, Origen emphasized the pedagogical and apologetic use of philosophy; by means of Greek categories, the full truth of Scripture, in opposition or agreement with philosophy, could be made comprehensible to non-believers or to educated Christians.[12] However, since he focused on Jesus as the Logos, his method probably appealed to already committed Christians more than to enquiring pagans. While his student, Gregory Thaumaturgus, marvelled at Origen's freedom and intimacy with philosophical and divine teachings, another listener, Porphyry, scored his methods as arbitrary and incoherent.[13] Origen himself acknowledged that the use of philosophy in spiritual life was a subtle and potentially dangerous undertaking, relying not on an uncritical belief in universal truth, but rather on the potential of an instructed Christian to sort out and apply the useful aspects of secular learning for spiritual

[10] His *Hom. Gen.* 7. 2, 12. 4, 13. 3, and *Hom. Lev.* 1. 1 are cited as examples by Nautin, *Origène*, 405 n. 114.

[11] Nautin, *Origène*, 441.

[12] *HE* 6. 19. 11–13.

[13] Contrast Gregory, *Pan. Or.* 7. 93f. (*SC* 148. 134f.), with Porphyry in *HE* 6. 19. 4–8.

progress.[14] With confidence in both the ability of the individual and the revelation of God through the Logos, the teacher or exegete merely laid the groundwork for the gradual, deepening communion of each person with God.[15]

The question persists as to whether Origen was able to fulfil his theological vision under the weight of contemporary Platonism. Recent studies of his work have revealed not only his critical detachment, but also his original and subtle use of philosophical categories to express Christian concerns; as J. M. Rist has commented, 'Ars est celare artem.'[16] Increased knowledge of ancient religion and philosophy seems to delineate rather than circumscribe his theological intentions. Origen drew on philosophical language, yet within his theodicy attempted to make it serve distinctively Christian ends. Equally important, these ends must be recognized as those of the third century and of a largely independent teacher, and should not be attributed to the philosophy on which he was drawing solely because they are by later standards unorthodox. For example, his use of hierarchical scriptural texts such as Mark 10: 18 should be taken seriously as representing theological or scriptural axioms, rather than as mere proof texts for Platonic cosmological assumptions. This may be helped by locating Origen's doctrinal discussions more clearly in their specific religious contexts. Whereas all his works portray his synthesis of metaphysical and religious concerns, those such as *On Prayer* or the *Dialogue with Heraclides* reveal more clearly his inclination to discuss the philosophical considerations as the background to practical spiritual and theological problems.[17]

Whether or not Origen may be called systematic in the strictest sense, it is clear from *On First Principles* and the function of his theological language generally that he deliberately set out an organized metaphysical scheme within which to explore

[14] *Letter of Origen to Gregory*, 3. 2 (*SC* 148. 190f.).

[15] K. Torjesen, 'Origen's Interpretation of the Psalms', *SP* 17.2 (1982), 957; idem, *Hermeneutical Procedure and Theological Method in Origen's Exegesis* (Berlin, 1985).

[16] J. M. Rist, 'Beyond Stoic and Platonist', in Blume and Mann (eds.), *Platonismus und Christentum*, 238; P. M. O'Cleirigh, 'The Meaning of Dogma in Origen', in Sanders (ed.), *Jewish and Christian Self-Definition*, 203; R. M. Berchman, *From Philo to Origen* (Chico, Calif., 1984), 113f.

[17] See his discussion of *ousia* in *Or.* 27. 8 (*GCS* 3. 367–8). Harl notes a development from speculation to spiritual insight and mystical exercise in Origen's work in *Origène et la fonction révélatrice*, 362.

contemporary problems of the relation of God, Christ, and the world.[18] The structure of the work has obvious sources in philosophical thought, but it was new as a Christian attempt to present the science of God as a coherent whole. Regular teaching would tend to fix such a model, but would equally require a degree of flexibility, which is also characteristic of Origen's work. Origen may best be described not as systematic, but as ordering his theology and exegesis on the basis of certain axiomatic principles from philosophy and theology.[19] By establishing the sources and function of these principles, one may hope to evaluate his intentions correctly.

However, notwithstanding these principles, Origen often remained open and searching with reference to difficult issues, and cannot always be pressed without fear of distortion. In his lifetime his Gnostic opponent Candidus was able to create a turmoil by turning a reflection on the devil from hypothesis to certainty; Origen's point was to uphold the freedom of the will, rather than teach specifically about the ultimate salvation of the devil.[20] At times, therefore, he presented a number of alternative answers to speculative questions. His controversial doctrine of pre-existence was similarly pressed from a deliberate choice of a possible theory to a Platonizing error by his critics.[21] The contrast between his exploratory style and simple doctrinal affirmations is illustrated in the *Dialogue with Heraclides*; Origen's careful teasing out of the nature of the soul, often impossible to answer and yet requiring a helpful pointer, was reduced to a simple answer by his episcopal listeners: 'Brother Origen teaches that the soul is immortal.'[22] Not surprisingly, the

[18] On the structure and purpose of *De Principiis*, see H. Crouzel, 'Origène est il un systematique?', in *Origène et la philosophie* (Paris, 1962), 180f.; F. H. Kettler, *Der ursprüngliche Sinn der Dogmatik des Origenes* (Berlin, 1966), 1–5; M. Harl, 'Structure et cohérence du Peri Archon', in H. Crouzel, G. Lomiento, and J. Rius-Camps (eds.), *Origeniana* (Bari, 1975), 11–32; H. Crouzel and M. Simonetti (eds.), *Traité des Principes* (Paris, 1978–84).

[19] Harl, *Origène et la fonction révélatrice*, 367.

[20] 'Asserit Candidus diabolum pessimae esse naturae et quae salvari numquam possit. Contra hoc recte Origenes respondit non eum periturae esse substantia sed voluntate propria corruisse et posse salvari' (*Con Ruf.* 2. 19, cited in Nautin, *Origène*, 169).

[21] *Princ.* praef. 5 (*SC* 252. 84). On other multiple answers as matters of discussion, see *Princ.* 2. 3. 7 (*SC* 252. 270–4), 2. 8. 4 (*SC* 252. 348), 2. 8. 5 (*SC* 252. 350–2).

[22] *Dial.* (ed. Scherer, 166).

hesitations and accommodations which were essential to his teaching were often overlooked in favour of the lovely or repelling coherence of his theology. His pedagogical ambiguity in combination with his clear axiomatic principles allowed later, seemingly consistent doctrines to be attributed to him.[23] His theology was in fact a network of theological teachings rather than a body of systematic principles. Thus the separate parts of the theological model have a clear integrity and importance, yet they are revealed most clearly in their relationship to one another, and function in expressing his distinctive world-view.

Origen's passionate search for conceptual clarity and spiritual advance was grounded in his devotion to study as a means of communication and relation to the Logos.[24] His masterful use of Middle Platonic ontology and Stoic psychology, his enquiries into Hellenistic Jewish exegesis, and his relentless repudiation of Gnostic election theology were all ordered by his optimistic belief in the accommodation of the Logos to the mind and life of the individual person. If his spirituality was highly intellectual, it was not mystical in the strictest sense of the term.[25] Rather it consisted in the intense, disciplined search for God through the movement of the mind as led by the Logos. Thus, Origen urged people to be bold and free with the faith and consider many options, any one of which may turn out to be the way in which the Logos has chosen to bring that particular person nearer to divine truth.[26] Teachers were thus spiritual guides who attempted to destroy illusions and themselves laboured so as to grow in greater intimacy with the Logos.[27]

If Origen's Logocentric method had an obvious strength and appeal in an era of small urban study groups and independent teachers, it became a liability in the more structured authority of the later Christian community. Origen was undoubtedly typical of many early Christian teachers in his intellectual creativity,

[23] A. Guillaumont, 'Évagre et les anathématismes antiorigénistes de 553', *SP* 3 (1961), 213–26; F. Refoule, 'La Christologie d'Évagre et l'origénisme', *Orientalia Christiana Periodica*, 27 (1961), 221–66.

[24] Torjesen, *Hermeneutical Procedure*; C. W. Macleod, 'Allegory and Mysticism in Origen and Gregory of Nyssa', *JTS* NS 22 (1971), 368–9.

[25] A. Louth, *The Origins of the Christian Mystical Tradition from Plato to Denys* (Oxford, 1981), 71; Berner, *Origenes*, 68–84.

[26] *Can.* 3. 8 (ed. Lawson, 197); *Princ.* 2. 9. 4 (*SC* 252. 360).

[27] *Hom. 1 Can.* (ed. Lawson, 283); *Can.* 1 (ed. Lawson, 61); *Can.* 2. 5 (ed. Lawson, 138); *Can.* 3. 11 (ed. Lawson, 211–14).

spiritual intensity, and loyalty to the church. However, he later came to be regarded as someone whose speculative interests had given rise to many questions which were uncomfortable or superfluous to the larger community. His theological vision hardly survived his death in any complete way. In the eyes of contemporary, as well as later critics, Origen's teaching was weakened by his philosophical and subjective perspective. His focus on intellectual reflection ignored the problems of everyday life and unruly passions;[28] and his unorthodox speculations made even his spiritual insights suspect.[29] Ironically, the persisting influence of his theology lay precisely in his optimistic belief in the power of the Logos, accessible through speculation and individual study. Like the ancient *polis*, the cosmos was a diverse, but ultimately unified and sociable place for the bold and educated citizen: theology is the wine which makes us happy, heaven the place where we will learn why and how each star is placed and converse daily with the Logos. The best conversation revolves around Scripture and the search for truth; whatever one's circumstances, mobility and reunion with God are possible.[30] Origen's eager questions and passionate defence of the self-determination of the individual soul reflect a Christianity of small groups and spiritual guides set against a suspicious and violent, but largely stable, urban society. Through the gracious power of the Logos, all people are able to grasp the mysteries of transcendent philosophy; if Origen grew impatient with the simple, he did not deny them the kingdom of heaven.[31] As a philosopher and a Christian, he defended cosmologically the individual journeys and meanderings of all people towards God guided by the individual care and instruction of the incarnate Logos.

[28] Harl, *Origène et la fonction révélatrice*, 360; Stead, 'Freedom of Will', 252.

[29] For early monastic criticism of Origen, see B. Ward (tr.), *The Sayings of the Desert Fathers* (Kalamazoo, Mich., 1975), 103; *Vita Pachomii*, 31 (56), ed. and tr. A. Athanassakis, *The Life of Pachomius* (*Vita Prima Graeca*) (Missoula, Mont., 1975), 42.

[30] *Jo.* 1. 208 (*SC* 120. 162); *Princ.* 2. 11. 4–5 (*SC* 252. 400–6).

[31] On love and endurance, see *Prol. Cant.* 2 (ed. Lawson, 38); also Brown, *Making of Late Antiquity*, 70–1 'For all his sharp perfectionism, Origen's view of man was one that placed the main emphasis on the permanent resources of the individual identity, and on the slow, sure processes by which this identity unfolded ever higher and higher potentialities.' See also Gunnar af Hällström, *Fides simpliciorum according to Origen of Alexandria* (Helsinki, 1984).

ORIGEN AS THEOLOGIAN

Divine nature and will

As a teacher and a theologian, Origen wished, for intellectual and religious satisfaction, to describe the saving God revealed in Jesus and in Scripture. For this purpose neither the absolute transcendence affirmed by the Platonists nor the literalism of some Jews and Christians was worthy of this at once powerful and loving deity.[32] As seen in *On First Principles* and also, less formally, throughout his commentaries and homilies, Origen's discussion of divine nature and will was not merely a speculative exercise, but a religious necessity to guard against the besetting spiritual distractions of the third century: fatalism, despair, superstition, and idolatry. He attempted to blend philosophical expressions of transcendence and goodness with religious affirmations of God's direct intervention in the structure of the world. The result was a dynamic ontology which has been described as 'relational', or 'participatory', since all existence is defined as good by its relationship to God's being and will.[33] Against contemporary fatalism, essential goodness, together with power in the divine will, overcame material contingency, diversity, and weakness.

Like Justin and Irenaeus, Origen accepted and used Middle Platonic formulas of transcendence as definitions of divine nature. God as *νοῦς* or *ἀγένητος* was the source of all things, and was wholly comprehensible only to himself, transcending all attributes, perhaps even being itself.[34] God was simple (*ἁπλοῦς*), one (*ἕν*, *μονάς*), incorporeal (*ἀσώματος*), unchangeable (*ἄτρεπτος*), and invisible (*ἀόρατος*).[35] As Origen stated in his

[32] Cf. *Cel.* 7. 43 (*SC* 150. 114–16).

[33] H. Holz, 'Über den Begriff des Willens und der Freiheit bei Origenes', *Neue Zeitschrift für systematische Theologie*, 12 (1970), 83.

[34] For God as *νοῦς*, see *Princ.* 1. 1. 6 (*SC* 252. 100); *Cel.* 7. 38 (*SC* 150. 100). For God as *ἀγένητος*, see *Jo.* 2. 2 (12) (*SC* 120. 214). On transcendence, see *Princ.* 1. 1. 6 (*SC* 252. 100–4); *Cel.* 7. 38 (*SC* 150. 100), 6. 62 (*SC* 147. 332–4). On the relation of God and being, see *Cel.* 7. 38 (*SC* 150. 100), 6. 64 (*SC* 147. 338–40); *Jo.* 13. 123 (*SC* 222. 94–6), 19. 37–8 (*SC* 290. 68); cf. Whittaker, '*Ἐπέκεινα νοῦ καὶ οὐσίας*'.

[35] For God as *ἁπλοῦς*, see *Jo.* 1. 20 (*SC* 120. 122); for *ἕν*, see *Princ.* 1. 1. 6 (*SC* 252. 100); *Jo.* 1. 119 (*SC* 120. 122); *Cel.* 7. 38 (*SC* 150. 100). For *μονάς*, see *Princ.* 1. 1. 6 (*SC* 252. 100), and cf. *Jo.* 1. 119 (*SC* 120. 122); *Hom. Gen.* 3. 2 (*SC* 7. 118). For God as *ἀσώματος*, see *Princ.* 1. 1. 1–8 (*SC* 252. 90–110); *Jo.* 13. 123

preface to *On First Principles*, he has assimilated the biblical predications of God (fire, light, spirit) to the philosophical definitions, of the incorporeal. Rather than definitions of divine substance, these were descriptions of God's relationship to creatures as power to cleanse, enlighten, and inspire.[36] Like Philo, he used philosophical presuppositions to allegorize biblical anthropomorphic descriptions of divinity such as God walking.[37]

However, like earlier theologians, Origen accepted philosophical divine transcendence while arguing that God is the active source and power within all existence. Most directly this was a refutation of Gnostic and Marcionite dualism, as well as contemporary fatalism, in that it describes God's ongoing, direct connection with the creation by means of nature and will. Thus, Origen used the common Platonic idea of participation not as an abstract relation to a constitutive Form, but rather to express a state or condition of individual relationship to an active, intentional divine being.[38] In this construction, creatures are profoundly linked to God by both their nature and their activity: Origen claimed that everything received God's spirit and seed by participation.[39] This was not to be taken literally; he explicitly denied that God was like a pantheistic deity or spirit in any material sense.[40] The significance of participation was not to prove that the cosmos was ὁμοούσιος with God, as in a Gnostic system, but rather to show that all that exists comes from the divine essence in an intentional way, and continues in a dynamic, enduring relationship with God who is the source and ground of existence. Created nature, therefore, only makes sense in light of its constitutive relationship with the Creator. In contrast to the Gnostic idea of creation, the very fact of existence in Origen's scheme ensures that nothing spiritual or material is separated from God. In a theological extension of the philosophical model, existence itself takes on the character of a personal relationship. In his *Commentary on Romans* Origen compares existence to

(*SC* 222. 96); *Cel.* 7. 38 (*SC* 150. 100); *Hom. Gen.* 3. 2 (*SC* 7. 114). As ἄτρεπτος, see *Cel.* 4. 14 (*SC* 136. 216) and *Or.* 24. 2 (*GCS* 3. 354). For God as ἀόρατος, see *Princ.* 1. 1. 8 (*SC* 252. 106), 2. 4. 3 (*SC* 252. 286); *Cel.* 6. 64 (*SC* 147. 340).

[36] Deut. 4: 24, John 4: 24, and 1 John 1: 5 are cited in *Princ.* 1. 1. 1 (*SC* 252. 91 f.); cf. *Jo.* 13. 129–40 (*SC* 222. 98–106).

[37] *Hom. Gen.* 1. 13 (*GCS* 29. 15), 3. 1. (*GCS* 29. 39); *Or.* 23 (*GCS* 3. 349–53).

[38] See Stead, *Divine Substance*, 141.

[39] *Princ.* 1. 3. 6–8 (*SC* 252. 154–62).

[40] *Cel.* 6. 71 (*SC* 147. 356–60).

grace as a universal relationship and gift of God: 'Grace is an act of sheer kindness, no reward. . . . Just as for example the fact we exist cannot be a reward of our gifts, but is due to the grace of our creator.'[41]

An example of how he uses Scripture and philosophy to express this theology of participation and creation may be seen in his exegesis of Exodus 3: 14. Origen referred to ὁ ὤν in Exodus 3: 14 as a name of God.[42] In his early commentary on Psalms, used by Philo and possibly Numenius, ὁ ὤν was taken as the name of God given to Moses in the Septuagint, a name which indicates that God is the source of all existence. Origen most often used it to describe the relation of being which God has bestowed on humanity.[43] In the early works *On First Principles* and the *Commentary on John*, Exodus 3: 14 is cited as expressing God's free, good gift of existence to all creatures.[44] Used in conjunction with Mark 10: 18, Exodus 3: 14 defines evil as non-being, rather than a substance; evil is therefore separation from God, rather than an independent force.[45] In his later commentaries, Origen continues to link grace and participation with reference to Exodus 3: 14; in his Pauline commentaries in particular, he refutes Gnostic election theology by underlining God's continual and chosen relationship with all existence.[46] However, this relationship entails shared existence, not shared essence. As the source of everything, divine being is unchangeable and unlike contingent, hence accidental being; shared existence is not shared nature. Even more explicitly, as ὁ ὤν, God is the source of all existence by will, not essence.[47]

This relationship between the divine essence of the Creator and the contingent creation may be further clarified with reference to Origen's exegesis of goodness. Like most of his contemporaries, he accepted the transcendent goodness of God as axiomatic, if

[41] *Com. Rom.* 4. 5, cited in B. Drewery, *Origen and the Doctrine of Grace* (London, 1960), 19.

[42] Cf. P. Nautin, '"Je suis celui qui est" (Exode 3. 14) dans la théologie d'Origène', in idem, *Dieu et l'être* (Paris, 1978), 109–19.

[43] J. Whittaker, 'Moses Atticizing', *Phoenix* 21 (1967), 196–201; cf. Stead, *Divine Substance*, 106–7.

[44] *Princ.* 1. 3. 6 (*SC* 252. 154); *Jo.* 2. 96 (*SC* 120. 268).

[45] *Jo.* 2. 96 (*SC* 120. 268–70).

[46] *Com. Eph.* Fr. 2 (*JTS* 3 (1901–2), 235); *Com. Rom.* 25 (*JTS* 13 (1911–12), 361); *Or.* 24. 2 (*GCS* 3. 354); *Com. Mt.* 17. 36 (*GCS* 40. 700).

[47] *Hom. 1 Reg.* 1. 2 (*GCS* 33. 153).

not commonplace. His theological interest was sharpened, however, by polemics against the Marcionites and some Gnostics who denied the goodness of the Creator. In *On First Principles* and the *Commentary on John*, he uses Platonic as well as scriptural ideas to combat this idea. Origen describes God not only as the highest source of goodness, but also as the active creator of a good world who continues to give good things to all creatures.[48] In contrast to the essential and unchanging goodness of divine nature, creatures are only good accidentally; the link between the goodness of divine nature and the goodness of creatures is the derivative or image of divine goodness, which is the Logos.[49] This ontological hierarchy of existence and essence is translated in his theodicy into a dynamic cosmology of individual progress toward salvation. In eternal perfection and freedom, God who is always doing good is the archetype for human action and aspiration.[50] Thus, the goodness of Christ's acts reveals his divine nature, and creatures become closer to God by means of consistently good actions.[51] In several works Origen structures this dynamic hierarchy of goodness by means of biblical verses, including Mark 10: 18 and Wisdom 7: 28.[52] Whereas the philosophical contrast between essential and accidental goodness orders the ontological relation, the action of the will underlies the soteriological relation between God and humanity. Even if essentially accidental, creatures are related to God by a shared existence; but the nature of that existence, whether original or derived, preserves the distinction between divine and human nature.

Even within the Trinity divine substance is defined by qualities and activity in relation to the Father, the source of all being. On the one hand, Origen describes the Son and the Spirit as unquestionably divine, incorporeal, and essentially good.[53] On the other hand, because the kind of nature is determined by the order of procession from the Father, both the Son and

[48] *Cel.* 3. 28 (*SC* 136. 66–8); *Jo.* 1. 25 (*SC* 120. 253–4).

[49] Cf. *Princ.* 1. 2. 13 (*SC* 252. 140), 1. 2. 4 (*SC* 252. 116), 2. 9. 2 (*SC* 252. 354); *Cel.* 6. 44 (*SC* 147. 286). See also Berchman, *Philo to Origen*, 154.

[50] *Cel.* 4. 28 (*SC* 136. 250); *Hom. Jer.* 9. 4 (*SC* 232. 392).

[51] *Cel.* 7. 33 (*GCS* 150. 184).

[52] Cf. *Princ.* 1. 2. 13 (*SC* 252. 140–2); *Com. Mt.* 15. 10 (*GCS* 40. 374–6).

[53] *Princ.* 1. 2. 13 (*SC* 252. 140–1), 1. 6. 4 (*SC* 252. 204–6), 1. 6. 2 (*SC* 252. 198), 1. 8. 3 (*SC* 252. 226–8), 2. 2. 2 (*SC* 262. 246), 4. 3. 15 (*SC* 268. 396).

the Spirit are derivative and can thus be called created, and the Spirit who comes through the Son is sometimes described as the lowest element.[54] Equally important, he distinguishes the divine essence by function or action. Against modalists, Origen defines the function as essential to the person, not as a temporary economic phase.[55] On account of this synthesis of essence and activity, he may refer to one divinity or will or *οὐσία* in the Trinity, yet also insist on their separation *κατ' οὐσίαν*.[56] Nature is shared, yet distinguished by activity or relationship to the Father, within the Trinity as well as in creation. Even though his insistence on the uniqueness and eternity of the relation of the Father, Son, and Spirit precludes the extreme subordinationism sometimes attributed to him, the core of their unity remains the relation to the Father, the source of existence.

Because of the importance of the existential relation to the Father through individual activity, many scholars have pointed to the dynamic quality of Origen's ontology. To underline the importance of individuation by relation, F. Kettler describes Origen's ontology as existential: a process of being, knowing, and communicating.[57] H. Holtz follows E. Benz in using more voluntarist categories to describe the movement: will is not accidental or exclusive of substance, but is part of the entity constituting the category of person in Origen.[58] J. Rius-Camps and M. Simonetti have applied Stoic conceptions of *πνεῦμα*, the universal substratum of all things, which is differentiated only by quality, by way of analogy in an attempt to understand Origen's intentions. Rius-Camps literally equates Origen's use of *πνεῦμα*, taken from biblical texts, with a concept of God permeating all existence, whereas Simonetti takes more seriously Origen's repudiation of Stoic materialism, and defines the Trinity as a triad of individuals with

[54] *Princ.* praef. 4 (*SC* 252. 80–2); *Jo.* 13. 151–3 (*SC* 222. 112–15).

[55] *Princ.* 1. 2. 5–6 (*SC* 252. 118–20). See H. Crouzel, 'Les Personnes de la Trinité sont-elles de puissance inégale selon Origène, Peri Archon 1, 3, 5–8?', *Gregorianum*, 57 (1976), 109–25.

[56] *Cel.* 8. 12 (*SC* 150. 200); *Or.* 15. 1 (*GCS* 3. 334).

[57] F. Kettler, 'Die Ewigkeit der geistigen Schopfung nach Origenes', in M. Greschat and J. F. G. Goeters (eds.), *Reformation und Humanismus* (Witten, 1969), 272–97.

[58] Holtz, 'Über den Begriff des Willens', 69–71; E. Benz, *Marius Victorinus und die Entwicklung der abendlandischen Willensmetaphysik* (Stuttgart, 1932), 335.

shared power.[59] Within the triad, as well as in the created order, individuals have different qualities of being in relation to the Father, and hence different identities. Simonetti, Holtz, and, most recently, Berchman emphasize the unique dynamism of Origen's metaphysics, which consciously fuses essence and activity in order to define and link God and creation as well as Father, Son, and Spirit.

Divine will in Origen's theology must therefore be understood within this particular dynamic vision of nature and identity. Koch and Amand have rightly pointed out his dependence on philosophical discussions of providence and freedom.[60] Yet, as noted by Holtz, in many places Origen moves cautiously beyond the Platonic notion of will as an expression of rational nature to the theological expression of God's unhindered, personal purpose in the order of creation and in individual lives.[61] Although Origen did not analyse the specific problem of the relationship between divine will and divine nature in the extant texts, his descriptions of the activity of God reveal his struggle to balance consistent purpose with personal character. Like Irenaeus, he argues that to be good and just, God must be able to create as he chooses; otherwise the idea of divinity is inadequate, and the hope of Christians is in vain. The necessary corollary of divine goodness, therefore is divine power and intention. Will is often the linchpin in his descriptions of God's activity in creation and redemption: one cannot deny divine goodness or power, so is God unwilling? Divine will is both the immediate expression of God's nature and the interior desire of God: 'Mensura enim Dei caritatis haec sola est ut tandem, quantum ipse vult, diligatur; voluntas autem Dei eadem semper est nec unquam mutatur. Numquam ergo immutatio aliqua aut finis ullus in Dei "carite" recepitur' (*Cant.* 3 (*GCS* 33. 198)). However, if divine will is immutable in purpose, its operation is not automatic, but reflects God's personal relationship with the changing creation. In one

[59] J. Rius-Camps, *El Dinamismo trinitario en la divinizacion des los seres racionalis segun Origenes* (Rome, 1970), 89; M. Simonetti, 'Note sulla teologia trinitaria di Origene', *Vetera Christianorum*, 8 (1971), 277f. See also C. Blanc, 'Dieu est *pneuma*', *SP* 16.2 (1985), 224–41.

[60] Koch, *Pronoia und Paideusis*; Amand, *Fatalisme et liberté*, 297f. See also P. Kübel, *Schuld und Schicksal bei Origenes, Gnostikern und Platonikern* (Stuttgart, 1973), 84f.

[61] Holtz, 'Über den Begriff des Willens', 70.

passage Origen discusses βουλή and θέλημα in Ephesians 1; θέλημα is God's intention, which shares the immutability of his nature.[62] The intention of God may be distinct from God's actual response to events in the world:

Memento quia nihil sine providentia eius geri diximus; non, sine voluntate. Multo enim sine voluntate eius geruntur, nihil sine providentia. Providentia namque est qua procurat et dispersat et providet quae geruntur, voluntas vero est qua vult aliquid aut non vult. (*Hom. Gen.* 3. 2 (*SC* 7. 116))

To safeguard both the power and the goodness of divine will, God is said to be not merely a stationary, or immoveable, first principle, but an active, responsive agent. Will (*voluntas*), as God's inner desire and the outward expression of his plan, is used as an analogy for the generation of the Son:

qui utique natus ex eo est velut quaedam voluntas eius ex mente procedens. Et ideo ego arbitror quod sufficere debeat voluntas patris ad subsistendum hoc, quod vult pater. Volens enim non alia via utitur, nisi quae consilio voluntas profertur. (*Princ.* 1. 2. 6 (*SC* 252. 122))

In this passage Origen asserts the absolute power of the divine will in the creation of anything, the Son or the cosmos, and the close relation between mind and will, which is in line with contemporary Stoicism, which assimilated the intellectual to the volitional process. Will is not only the actualization of rational reflection, but part of the reflection itself (*consilio voluntas*). It is in this sense that he refers to divine will as the linchpin in the creation or generation of wisdom:

For one would say either that God was unable to create [*non potuisse deum dicet generare*] wisdom before he created it, so that he brought into being at a later time what had not existed before, or that he could do so, and—what is forbidden to say about God—he refused to create it [*noluisse generare*]. Each of which appears senseless and impious to all, namely that either God brought about something he could do from something he could not or while he could do it, he pretended not to, and delayed doing so. (*Princ.* 1. 2. 2 (*SC* 252. 112–14))

Here he does not appeal to whether wisdom was part of a rational plan, but to whether having the power, God chose to beget wisdom. Once again, a trace of a more voluntaristic con-

[62] *Com. Eph.* (*JTS* 3 (1901–2), 241–2); *Com. Gen.* = *PE* 7. 20 (*GCS* 43. 1. 402).

cept of God is evident, although Origen teaches a kind of eternal choosing, rather than an arbitrary or temporal decision to preserve the immutability and goodness of divine nature and power; will reflects the exercise of the intention of God. In a similar way, he argues, that only God's will bestows grace; yet he also assures his reader that God's will continually acts towards humanity.[63] Origen describes divine will, therefore, as God's inner desire, in order to affirm both ultimate power and faithfulness.

This concept of divine will is essential to Origen's description of creation. Like descriptions in the writings of other Christian and Jewish theologians, Origen's description may be distinguished from those of the Platonists and the Gnostics by his insistence on God's intention and power to enact his goodness and desire in material creation. God was neither limited in his design by unstable matter nor separated by inferior intermediaries. God created exactly as he wished:

> On the basis of divine scripture, we have frequently declared that God is the good, just, and omnipotent creator of all things. Thus, when in the beginning he created the things which he willed to create [*in principio creavit ea, quae creare voluit*], that is rational natures, he had no other means for creating, except himself, in other words his goodness. (*Princ.* 2. 9. 6 (*SC* 252. 364))

The theological issue in this particular passage is anti-Gnostic. God as good and just created all equal and alike, so that any diversity was a result of the Fall, not of predestination or chaotic matter.[64] In the *Commentary on Genesis* Origen denies that the will of God was unable to create as God chose; hence divine will established the qualities of the non-existent, rather than being limited by eternal matter.[65] In this passage the power of God's will is emphasized, as proper to his divinity as is goodness. The philosophical background of Origen's account is clear: like that of the Timaean demiurge, God's reason for creation was goodness, and his purpose was to establish rational order. Yet, the theological point goes further: as God's power is unlimited, his actual good intention is expressed in the very nature of existence.

[63] *Cant.* 3 (tr. Lawson, 198). [64] *Princ.* 3. 3. 5 (*SC* 268. 194–6).

[65] *Com. Gen.* = *PE* 7. 20 (*GCS* 43. 1. 402): '*Ζητητέον πρὸς αὐτὸν περὶ δυνάμεως θεοῦ. εἰ θελήσας ὑποστῆσαι ὅ τι βούλεται ὁ θεός, τῆς θελήσεως αὐτοῦ οὐκ ἀπορουμένης οὐδὲ ἀτονούσης, οὐ δύναται ὑποστῆσαι ὅ βούλεται.*

Matter is not in opposition to the nature of God; nor does it limit divine power.[66] Hence any separation or limit in material existence is a result of moral imperfection.

Because divine power or will is shaped by divine nature, including the fact that it is eternal, Origen taught a doctrine of eternal creation. His intention was not to deny a distinction between Creator and creation, as is clear from his repeated defence of *creatio ex nihilo*, or to put forward an unqualified concept of eternal causality. He suggests that creation was eternally prefigured in Wisdom; hence God's eternal creating might be similar to divine foreknowledge: God foresees creation, but does not bring it into actuality immediately.[67] However, the deeper issue is the congruence of divine will and divine nature. As Creator and Father, God is always active in exercising benevolence:

> For it is at once senseless and impious to think that at any point or time God's powers were at rest . . . and therefore if no precise point can be supposed when this power was not doing good, there were always things for which he did good, namely his creatures, and doing good appropriately and justly in these things, he distributed his favours with the aid of providence . . . so we say neither that God's works are unbegotten and co-eternal with God nor that he was transformed to doing good having done nothing good before. (*Princ.* 1. 4. 3–5 (*SC* 252. 168–72))

The perfect congruence between eternal divine nature and will thus results in God's continuous salvific and creative action. If the description of the source of action is reminiscent of Platonic images of creation, especially in the denial of a change in God, the affirmation of an eternal desire and power to be doing good in creation has become an issue of piety. Origen underlines the titles of Creator and Father in order to express not only divine power and divine transcendence, but divine goodness as unhindered, active care, if not love of creation. The distinction between Origen's theological account and contemporary philosophical accounts rests on the degree of power affirmed and its reflection of the inner character of God.

As noted above, like contemporary Jewish and Christian

[66] *Princ.* 1. 4. 5 (*SC* 252. 172), 2. 1. 4 (*SC* 252. 242–3), 3. 4. 5 (*SC* 268. 214–15).

[67] *Princ.* 1. 4. 3–5 (*SC* 252. 168–72); cf. *Princ.* 1. 2. 2 (*SC* 252. 112–14).

theologians, he taught that God as Creator may transform matter and material existence as he chooses. God is above natural law, yet works through nature to lift human expectations to the level of divine nature.[68] Thus, as Creator, God may change a body into anything he wishes; divine will may sustain it beyond corruption. To a Platonist such as Galen or Celsus, the Christian claim that God may raise corruptible bodies was absurd; and Origen, as a theologian, took issue with Christians who relied on simple appeals to divine will alone to solve the religious paradoxes of creation and resurrection. He combined Stoic physics with an appeal to divine power so as to give a reasonable explanation of God's power in resurrection.[69] He did not intend to destroy the religious wonder of God's activity, but rather to make it intelligible within present reality, a reality entirely created and ruled by God. God's will thus shaped the whole of reality, material, intelligible, and spiritual. In a sense Origen's emphasis on God's power as above natural law echoes the Irenaean statement of God's will as the substance of all things. Although he assumes a separation between created and uncreated being as eternal and contingent, yet, by appeal to divine power operative at all levels of being, he is able to claim that certain qualities of being, like existence itself, may be shared, if in a different manner; thus created bodies can be transformed into incorruptible spiritual bodies. Although he accepted the Platonic hierarchy of intelligible and sensible worlds, he reduced the tension between them by his emphasis on the relationship between different manners of being linked by God's pervading will and power. No part of existence by either nature or will is far from God the Creator.

Theologically, Origen's use of philosophical notions of providence, against Gnostic election theology or contemporary fatalism, was essential to his Christian Platonic ontology. The shape of the universe is due largely to the interaction between divine will and nature and created free will, rather than an opposition of eternal and contingent nature. The error of the Marcionites was to separate God's nature and will, and thereby limit the hope of salvation; the error of the Platonists was to limit the

[68] *Cel.* 5. 24 (*SC* 147. 72–4).

[69] *Cel.* 3. 70 (*SC* 136. 158–60), 5. 23 (*SC* 147. 70f.); cf. H. Chadwick, 'Origen, Celsus and the Resurrection of the Body', *HTR* 41 (1948), 84; *Princ.* 3. 6. 4 (*SC* 268. 242); *Cel.* 7. 32 (*SC* 150. 84–8).

power of divine will.[70] The goodness of divine will was therefore axiomatic; God would not do the absurd or the shameful. If one has the misfortune to be a woman [*sic*] or a wild beast, it is not due to the stars, but rather to one's own desire, which God has foreseen rather than predestined; thus, God was not the author of Judas's betrayal of Jesus, even if he knew what actions would follow from Judas's fault.[71] Origen's defence of human freedom and divine justice was so strong that he even asserted that God foresaw rather than predestined the righteousness of the faithfulness of Jesus.[72] The interaction between divine will and human will rests on God's infinite goodness and human freedom, which he refuses to compromise on either side in light of Gnostic teaching of election. Thus, his exegesis of Romans rests on an expectation of universalism: God calls all and thereby justifies all.[73] Relying on the stability of this continual good will, a Christian should pray for the proper blessing.[74]

On account of Origen's acceptance of the eternal good will of God, some interpreters have seen divine love reduced by him to a static perfection. For example, Origen argues that God did not harden Pharaoh's heart, but that, instead, Pharaoh's refusal of God's long-suffering grace hardened his own heart.[75] The theological tension once again lies in Origen's attempt to blend a philosophical coincidence of nature and will with a religious affirmation of personal power expressed at all levels of being. While on the one hand he defends the consistency of divine power and love, on the other, he attempts to give it an individual expression in the circumstances of spiritual life. Thus, in his homilies and commentaries he gives a more dynamic and individualistic reading of divine will: God sets life before all creatures, rather than giving it to them; God rules rather than tyrannizes.[76] Divine will is therefore not an overruling force which overcomes human freedom, but a personal energy active in relation to particular events and actions. Thus, God brings human actions to fulfilment, and in answering prayers takes one beyond the

[70] *Princ.* 2. 9. 5 (*SC* 252. 360–1).
[71] *Com. Gen* = *PE* 6. 11 (*GCS* 43. 1. 351–3).
[72] *Com. Rom.* 2 = *Philocalia*, 25. 2 (ed. Robinson, 227–8). [73] Ibid. 227.
[74] *Or.* 6. 4 (*GCS* 3. 313–14).
[75] See W. J. P. Boyd, 'Origen on Pharaoh's Hardened Heart', *SP* 7 (1963), 434–42.
[76] *Dial.* (ed. Scherer, 172); *Hom. Jer.* 20. 2 (*SC* 238. 256).

original request to a deeper communion with God.[77] Indeed, God gives himself in these very gifts.[78] The whole shape of existence is therefore an engagement between human freedom and the saving will of God.

Origen's description of divine will and nature reflects the tension for a Christian in philosophical and theological concerns. As the divine nature is eternal, God's desire and actions must be equally so. The question of any temporal decision in creation must therefore be dismissed as impious, since God's will coincides with his nature. The obvious implication may be that God's will is an automatic or rational consequence of his nature. However, Origen's descriptions of the operation of divine will, shaped by biblical salvation history and the role of Christ, reflect a more voluntarist contour to his philosophical vocabulary. God's desire (*θελημα*) is the linchpin between his nature and power, so that the will (*βουλη*) of God operates in no impersonal or abstract way in creation and redemption. Origen's God is never beyond love for the individual or intervention; divine will is not merely for the good, but to increase the goodness of creation by specific actions. God as good and active, therefore, does not radiate goodness or rationality impersonally, but intentionally creates a free world to share divine life. Origen's dynamic cosmology thus highlights both the essential goodness and the saving will of God, which bridges the gap between eternal and contingent being. Creatures as contingent share existence with God by his will; and, as existent they may, however imperfectly, share the goodness of divine nature. This is grace: they exist to be drawn into intimacy with God. Confidence in divine power and active love is anchored ultimately in the very structure of eternal being and created existence.

Human nature and will

In his early *Commentary on Genesis* Origen notes that those who are supposed to have faith are distracted by doubt about necessity: 'Predestination gives glory to God, but destroys our free-

[77] *Or.* 29. 14 (*GCS* 3. 389); *Princ.* 3. 1. 9 (10) (*SC* 268. 56f.), 1. 8. 4 (*SC* 252. 228f.); cf. I. T. Holdcroft, 'The Parables of the Pounds and Origen's Doctrine of Grace', *JTS*, NS 24 (1983), 504.

[78] *Or.* 1. 1 (*GCS* 3. 297–8), God draws as well as invites (*Hom. Num.* 20. 3).

dom.'[79] A defence of free will is necessary, not only to safeguard God's justice, but also to preserve the sense of the entire historical economy of Christ's suffering and death, the Law and the prophets, the labours to establish the Church, and the efficacy of prayer. If Origen is to reject divine election, he must construct a coherent account of human freedom, which, in response to the alleged fatalism of the Gnostics, he accepts as a teaching of the apostles.[80] The degree of his insistence on human freedom was controversial, from his early work *On Nature* in which he claimed that the devil was evil by will not nature.[81] Ironically, some later critics of his theology have accused Origen of a philosophical moralism; others have perceived a Platonic account which ultimately makes Origen's anthropology essentially intellectual rather than volitional.[82]

Origen's controversial teachings about eternal creation, the pre-existence of souls, the incorporeal nature of the soul, and the relation of will and intellect were each introduced to lend coherence to certain other ideas, and must be examined in relation to one another as well as by comparison with philosophy. They were often discussed on occasions when he consciously proposed a speculative answer to a question, since tradition had no specific teaching. For example, Origen's idea of eternal creation is introduced in relation to the eternal will of God, not with regard to the nature of creatures; as seen above, God's will as always active would always have created. Creatures, as products of eternal will, are not ὁμοούσιος or eternal, as a Gnostic or a Platonist might teach. Origen denies that creatures are immutable or essentially good; instead, he defines them as existing in a contingent relation to the Father through the Logos. He explicitly denies that creatures are eternal or share God's essence in any literal sense.[83] In contrast to the Father, they have been created. The crux of the notion of eternal creation is God's desire for creation

[79] *Com. Gen.* = *PE* 6. 11 (*GCS* 43. 1. 350).

[80] *Princ.* praef. 5 (*SC* 252. 82); *Jo.* 32. 16 (*GCS* 10. 451).

[81] *Con Ruf.* 2. 19, cited in Nautin, *Origène*, 169.

[82] Divinization is described as wholly intellectual in J. Dupuis, '*L'Esprit de l'homme*' (Bruges, 1967), 161 n. 1, and in B. D. Jackson, 'Sources of Origen's Doctrine of Freedom', *CH* 35 (1966), 16 n. 27. Crouzel also inclines towards this view; cf. 'L'Anthropologie d'Origène dans la perspective du combat spirituel', *Revue d'ascétique et de mystique*, 31 (1955), 375. On Origen's philosophical moralism as unhindered intellectual ability, see Dihle, *Concept of Will*, 111–12.

[83] *Jo.* 13. 149 (*SC* 222. 112).

rather than the quality of human nature; only in the sense of being in the desire or wisdom of God are creatures eternal.

In the same way Origen's concept of the pre-existence of souls seems to be a way of understanding the inequality of present existence and the collective effect of the Fall, rather than an assertion of the soul's immortality or a previous spiritual life. In his introduction to *On First Principles*, Origen notes that the Church teaches that human free will and the fall of the angels are the sources of evil, but gives no fuller explanation; he therefore chooses a Platonic account as most useful in refuting any sense of determinism in one's present life.[84] Theologically, Origen denies that anyone is destined to a certain relationship with God by nature, insisting that all are created in equal and free relationship to God. From this original state, all have moved away from God, and one's present physical condition is a result of individual choice. There is no cosmic accident or divine predetermination in Origen's universe. The only necessity in God's creation is the outcome of the interaction between the divine will and individual human response.

This explanation of evil was controversial from its first inception, but it reflects several important aspects of Origen's cosmology and anthropology. First, all creatures are created equal and in direct relation to God; the divine creation is sovereign and intentional as far as all existence is concerned, so there are no mistakes and no lower orders as in Gnosticism. This original proximity to God resulted in all creatures being incorporeal, since Origen notes in Paul, as well as in Plato, that the original perfection was like the one to come, and hence was spiritual.[85] The only differences between persons comes about by individual choice, so variation in existence is moral. All souls, therefore, are responsible; they do not perish, but either return to God or receive punishment. Secondly, all souls except Jesus's have fallen away from God's presence, even if to greater or lesser degrees. The Fall, if individual in degree, is collective in outcome, leaving the entire cosmos groaning for liberation.[86] Using both Platonic and Pauline ideas, Origen describes material existence as a sign of estrangement from the perfect spiritual relationship with God.

[84] *Princ.* 2. 8. 4 (*SC* 252. 348). [85] *Princ.* 1. 6. 4 (*SC* 252. 204).

[86] Rom. 8: 22. *Cel.* 3. 62–3 (*SC* 136. 142–6); *Princ.* 1. 6. 2–3 (*SC* 252. 198f.), 1. 7. 5 (*SC* 252. 216–17), 2. 9. 2 (*SC* 252. 354–6).

However, physical existence is not a punishment, but a means whereby creation may return to God. The concept of the pre-existence of free souls allows Origen to express the universality of salvation history on a cosmic scale.

Origen's view of the cause of the Fall reveals the tension between the Judaeo-Christian story, Platonic categories, and his own emphasis on free will. He agrees with Irenaeus that to some degree, estrangement from God may be inevitable, because of the instability of created natures. But he also assigns individual responsibility to the degree of withdrawal from the goodness of God. The cause of the Fall remains unclear, however. Satiety of knowledge has been suggested by the use of the term κόρος. Yet, in light of Origen's emphasis on will, and following Harl, the more consistent cause is not satiety, but a lapse of attention in the spiritual life. By wilful withdrawal or neglect, the soul interrupts the flow of power and goodness obtained by relationship to God.[87] In his account, Origen focused not on interrupted contemplation, but on a defence of the dynamic self-determination of creatures and the goodness and justice of God. All levels of existence, from that of angels to humans to demons, were determined not by divine will, but by the individual response of the creature to God; thus response created identity.

If Origen did not subscribe to Irenaeus's wholly positive evaluation of material existence, neither did he describe it as a prison, as Augustine later claimed.[88] Within the Platonic cosmology and Pauline eschatology which framed his thought, the material world represented a subjected, alienated existence apart from God. The spirit was the image of God which yearns for eventual spiritual reunion. Thus, Origen may say on the one hand that creatures are incorporeal, but on the other claim that they have never existed without bodies.[89] Relying on texts such as Philippians 3: 21 and Romans 8: 20–1, as well as on contemporary Platonism, Origen defines the physical body as a temporary instrument of the soul.[90] Like the creation of death as an economic tool in Irenaeus's theodicy, the varieties of material and

[87] *Cel.* 6. 45 (*SC* 147. 290); *Jo.* 13. 203–14 (*SC* 222. 144–8); *Princ.* 2. 9. 2 (*SC* 252. 354–6); cf. M. Harl, 'Recherches sur l'origénisme d'Origène', *SP* 8 (1966), 373–405.

[88] Augustine, *Civ. Dei*, 11. 23.

[89] *Princ.* 2. 2. 2 (*SC* 252. 246–8).

[90] *Princ.* 2. 1. 3 (*SC* 252. 238–40).

spiritual existence are described by Origen for the purpose of educating the fallen soul. Thus bodies will be transformed into spiritual bodies and not abandoned on the Last Day.[91] Against the Gnostics, Origen tried to refute the idea that the body was a sign of punishment or was at enmity with God. His argument rested on a belief in the justice and goodness of God, who has arranged all creation. Thus, the variety of the material world is not as such a sign of imperfection or a prison, but a reflection of the various means whereby God redeems individual creatures.[92] Although as mortal, it is a sign of estrangement from God, material existence is not a punishment, as Augustine suggested on the basis of his reading of Origen; demons did not have heavy bodies, but were dark and cold spirits.[93] Rather than using a crude dualism between spiritual and material, Origen tried to outline the relation of existence by adopting Stoic categories of various qualities of being. At times he seems, therefore, to relate the state of the body directly to the state of the soul.[94]

Origen defines created humanity as both responsible and frail, but fundamentally free. Created in the image of God, humanity is a *ψυχή* or *νοῦς*, or sometimes *πνεῦμα*.[95] Origen emphasizes the original perfection of creatures as created in communion with God; yet, at other times, he mentions the progress necessary for humanity to grow from the image to the likeness of God.[96] The spirit is the conscience, which contains the image of God; in the end, all will be transformed into spirit. The human being is there-

[91] On the final transformation of bodies, see *Princ.* 1. 6. 4 (*SC* 252. 204), 2. 10. 1–3 (*SC* 252. 374f.), 3. 6. 4 (*SC* 268. 243f.).

[92] This includes social and geographical diversity, as well as moral states (*Princ.* 2. 9. 3 (*SC* 252. 356–7), 2. 11. 6 (*SC* 252. 406–10)).

[93] By contrast with God, who is fire and light (*Princ.* 2. 8. 3 (*SC* 252. 344–5)). Augustine's criticism of the material and spiritual in Origen's cosmology in *Civ. Dei*, 11. 23 (Why don't the demons have grosser bodies?) reflects a misunderstanding; Origen claimed that the wicked were dark and cold (*Princ.* 2. 10. 8 (*SC* 252. 392)).

[94] See D. Bostock, 'Quality and Corporeity in Origen', in H. Crouzel and A. Quacquarelli (eds.), *Origeniana Secunda* (Rome, 1980), 331f.; Brown, *Body and Society*, 165–8.

[95] *Princ.* 2. 8. 2–3 (*SC* 252. 340–8), 2. 11. 5 (*SC* 252. 404–6). On Origen's complex, inconsistent terminology, see Crouzel, 'L'Anthropologie d'Origène dans la perspective du combat spirituel', 364–85; idem, 'L'Anthropologie d'Origène: de l'arché au telos', in U. Bianchi and H. Crouzel (eds.), *Arché e Telos* (Milan, 1981), 36–43; cf. Dupuis, *'L'esprit de l'homme'*.

[96] *Sel. Ps.* 46, cited in Drewery, *Origen*, 50.

fore tripartite, consisting of spirit, soul, and body. Located between the spirit and the body, the soul is the place of decision and reflection; it is what mediates between the will of the spirit and the will of the flesh.[97] Free will is not lost in the separation from God, even if it is hindered by physical existence and estrangement from spiritual reality. As in Justin and Irenaeus, changeable humanity is defined as *αὐτεξούσιος*, or self-moved.[98] Origen uses this term positively to underline the natural autonomy of humanity, although in discussions of choice, he uses the more common philosophical term *προαίρεσις*.[99] The fundamental sense of human life and nature taught by Origen was therefore one of motion, either toward God or away, a motion which determined the qualities of the individual.[100]

Origen's discussions of training the will, ethical behaviour, and steadiness of purpose which rely on a Platonic perspective, have sometimes been seen as referring to a lower step of spiritual progress, moral life being merely a purificatory step to mystical communion based on the stabilized intellect. This analysis may neglect the importance of freedom in Origen's cosmology; the innate human potential for self-movement, *αὐτεξούσιος*, lies at its heart, as the continuous response of humanity to God. Rather than a moral responsibility based on proper intellectual discernment, *προαίρεσις* expresses theologically the potential appropriation of divine goodness by creatures; what is God's by nature, humanity acquires by will. Thus, action is not simply obedience to an external law or recognition of certain ideas; it is the entire motion of interior life in imitation of God: by will created nature is permitted everything.[101] The kinship between God and humanity rests in the innate human desire for God and the free potential to form a relationship with him. As discussed above, the created nature of humanity is unstable; it is mutable and contingent. Apart from the given existence with God, its qualities are

97 *Princ.* 3. 4. 2 (*SC* 268. 206).

98 *Jo.* Fr. 43 (ed. Brooke, 2. 258); *Or.* 6. 1–2 (*GCS* 3. 311–12); 29. 13 (*GCS* 3. 387–8).

99 Cf. Rist, 'Prohairesis', 112.

100 *Princ.* 3. 3. 5 (*SC* 268. 194); *Or.* 29. 13 (*GCS* 3. 387–8). On the continued identity of the individual, see G. C. Stead, 'Individual Personality in Origen and the Cappadocian Fathers', in *Arché e Telos*, 185f., esp. discussion following by Crouzel and Stead on character and physicality, 192–6.

101 *Hom. Ezech.* 1. 3 (*GCS* 33. 326); *Or.* 26. 5 (*GCS* 3. 361).

accidental and are maintained only through continual response to God. In Origen's essentially relational view of the universe, the ontological contingency which makes for instability also bestows moral freedom; all creatures are defined by this assimilation of mutability and moral freedom. Through individual response, creatures arrange themselves in relation to divine nature. By nature—that is, the fact of created existence—all are related to God; yet the depth of response or participation lies in the will of the creature. Thus, *μετέχω* does not function as an abstract relation, but rather describes the response to God which by communion or intimacy with divine nature increases or decreases the quality of existence.[102] Against the Gnostics, Origen defines the true self as centred in the capability and responsibility of free action in accordance with divine will.

In the third part of *On First Principles* Origen outlines an essentially Stoic account of the relation between the will and the intellect in order to safeguard individual responsibility. The exercise of will consists in the selection of the proper impressions from which to make a judgement. Impressions induce instability and can be misleading, but it is the unwillingness to judge thoughts and the willingness to nurture false or harmful thoughts which result in error or sin. Given reason and power as a rational soul, the individual is therefore responsible for making clear decisions free of illusion.[103] Origen relies here on contemporary philosophical commonplaces, but he uses them to justify a volitional account of human relationship with God which goes beyond moral freedom to spiritual transformation. As in contemporary Stoicism, contemplation is practically the same as volition; attention is as important as knowing the correct object.[104] To become like the images that one contemplates requires the power of will, as well as knowledge. Discipline of the fallen will to consistent attention is essential to salvation.[105] The Logos therefore accommodates itself not only to the level of comprehen-

[102] *Princ.* 3. 1. 20 (19) (*SC* 268. 40).

[103] *Princ.* 3. 1. 4 (3) (*SC* 268. 26); *Or.* 7 (*GCS* 3. 315–16). J. M. Rist, 'The Greek and Latin texts in *De Principiis* Book III', in H. Crouzel (ed.), *Origeniana*, 97–111; Stead, 'Freedom of Will', 252.

[104] Volke, *L'Idée de volonté*, 191–3.

[105] *Jo.* 20. 7 (*GCS* 10. 334). Carelessness causes hardness of heart or lets in demons (*Princ.* 3. 1. 13 (*SC* 268. 78 f.), 3. 2. 2 (*SC* 268. 158 f.)); whereas gradual exercise ensures lasting effect (*Princ.* 3. 1. 13 (12) (*SC* 268. 78–80)).

sion, but also to each individual's preference for certain ideas; the innate power of response is essential to human spiritual growth.[106] Origen's spirituality does not merely involve cognition as the source of stability and union with God, but rather the motion of the entire soul which must direct its attention to God and continually choose to fix it on God. Thus it requires choice and motion as much as knowledge.

In his description of the Fall and spiritual progress, Origen links moral actions and qualities of being. Just as *τόνος* determines the qualities of all existence, so continual attention to God allows the character of the individual to be transformed.[107] This is the process which Origen describes in his account of the union between Jesus's soul and the Logos.[108] Origen also used 1 Corinthians 6: 17 to express the goal of human life as becoming one spirit with God; the process of achieving this was specifically related to will and goodness.[109] *πνεῦμα* as the final goal of creation is a common concept in Origen, but this end is not to be one substance with God, but to achieve spiritual union.[110] Equally important, this process of transformation is not automatic; it depends on the power of God to change physical qualities, rather than the ability to manipulate one's own body by individual will.

The use of *χωρέω* is also linked to moral relationship between humanity and divinity in a quasi-physical sense. In *On Prayer* and the *Commentary on John*, the term is used to express Christ's relation to believers and also to express the divinity of the Son:

> Every member of the church ought to pray to contain [*χωρέω*] the Father's will as Christ contained it, who came to do the will of his Father and completed [*τελειώσας*] it. For by being joined to him, one becomes one spirit [1 Cor. 6: 17] with him, and in this way contains the will; so that as it has been completed in heaven, so it may be completed on earth. (Or. 26. 3 (GCS 3. 360)[111]

[106] *Princ.* 3. 3. 4 (*SC* 268. 192f.); cf. *Princ.* 3. 1. 15 (*SC* 268. 88), 3. 1. 21 (*SC* 268. 128f.).

[107] *Or.* 6. 5 (*GCS* 3. 315); cf. H. Cornélis, 'Les Fondements cosmologiques de l'eschatologie d'Origène', *Revue des sciences philosophiques et théologiques*, 43 (1959), 32–80, and criticism of Bostock, 'Quality', 331.

[108] *Princ.* 2. 6. 6 (*SC* 252. 322); cf. *Princ.* 2. 6. 3 (*SC* 252. 314–18).

[109] *Or.* 26. 3 (*GCS* 3. 360).

[110] *Dial.* (ed. Scherer, 126).

[111] Cf. *Jo.* 13. 231 (*SC* 222. 156), 32. 8 (*GCS* 10. 438).

The sense in this passage is of a free response in imitation of Christ's obedience, so as to be filled with God's gifts of 'sonship', goodness, and divinity and to accomplish the divine plan. Within Origen's categories of dynamic being with varying qualities linked to activity, *χωρέω* conveys both moral conformity and the transformation of the soul in communion with God.

Because of Origen's description of the intellectual kinship of the rational soul to God, the role of the will in divinization and final union with God has sometimes been neglected. However, as outlined above, will in a Stoic sense is integral to knowledge in Origen's account, and is part of the interior motion of his cosmology. Knowledge of God is essential to divinization in Origen's work, as a biblical and philosophical proposition; however, knowledge and reception of the Father and his will require not only a purified mind, but an active and continual response.[112] If the Fall was occasioned by a lapse of attention, redemption lies in its being sustained. From this perspective, it is false to distinguish between morality and mysticism in relation to Origen.[113] In his discussion of Judas, he explains that an evil thought harboured in Judas's mind, which allowed other such thoughts to grow and mislead him; a pure will was then perverted by the devil because of initial human carelessness.[114]

The ultimate spiritual goal was not static intellectual contemplation, but a continual and active union with God: 'But we say that at some time the Logos will have remodelled every soul to his own perfection, whenever each individual by his own power (*ἐξούσια*) chooses what the Logos wills, and obtains the state he has chosen.'[115] Although Origen was highly optimistic about the potential of human beings, he also recognized the need for constant divine assistance, the unassisted will being incapable of consistent obedience.[116] This is due not merely to its changeable

[112] *Or.* 29. 15 (*GCS* 3. 390); *Princ.* 3. 1. 24 (*SC* 268. 148–50).

[113] Crouzel, 'L'Anthropologie d'Origène dans la perspective du combat spirituel', 375.

[114] On Judas, see *Jo.* 32. 19 (*GCS* 10. 458); also S. Laeuchi, 'Origen's Interpretation of Judas Iscariot', *CH* 22 (1953), 253–65.

[115] *Cel.* 8. 72 (*SC* 150. 340). Cf. *Cel.* 5. 21 (*SC* 147. 66), 6. 20 (*SC* 147. 228–9); *Princ.* 3. 3. 6–7 (*SC* 147. 196–8). Compare Stead's emphasis on the static moral ideal in, 'Freedom of Will', 252.

[116] *Cel.* 7. 33 (*SC* 150. 88–90), 7. 42 (*SC* 150. 114), 7. 44 (*SC* 116); *Princ.* 3. 1. 18 (19) (*SC* 268. 118); *Or.* 29. 19 (*GCS* 3. 392).

nature, but also because within Origen's theology will itself is not merely an undetermined power, but the expression of piety and growth toward new life. One does not become perfect by will alone; divinization consists not in habit, but rather will and love directed toward God.[117] The wavering soul is strengthened by each choice that moves it towards its proper goal; the power of willing increases as one grows in relation to the Logos.[118] Thus, Judas's failure was not his betrayal of Jesus. It was his despair. He doubted divine mercy, and therefore did not turn again to God to strengthen him.[119] There is also a warning against complacency in Origen's account of Judas: even in a disciple, small weaknesses lurk which have to be disciplined continually to avoid sin. Temptations are inevitable; but one should be not only disciplined, but open to growth in holiness and the need for repentance because of sin.[120] The soul in constant motion does not remain neutral, and the motives and choices of the interior life must be constantly directed towards God. Origen accepts and affirms the changeability of human nature as the key to repentance, and charts the necessity of steady progress towards the good via continual choices. The transformation into the likeness of God steadies, but this does not negate the action of the will, for the active will of humanity towards God symbolizes the love between them.

The final union of believers with God is thus natural and volitional. Created in the image of God, humans by discipline, endurance, and grace grow into the likeness of God; from the beginning of the world all are called to 'sonship'.[121] The result is determined not by nature, but by God's will; *contra* Gnostic theology, Christians are 'sons' by will, not nature. However, all those are 'begotten' of God who do good works; they are eternally begotten in continued good works, and the spirit of 'sonship' remains within them. The slave becomes a disciple, then a

[117] *Com. Rom.* (ed. Scherer, 166–7).

[118] *Hom. Jer.* 14. 10 (*SC* 238. 86); *Cel.* 6. 20 (*SC* 147. 228); *Or.* 25.2 (*GCS* 3. 357–8).

[119] *Jo.* 32. 19 (*GCS* 10. 458); *Mt. Ser.* 117: the devil implanted a sadness unto death, cited in Laeuchi, 'Judas Iscariot', 259.

[120] On temptations, see *Princ.* 3. 2. 6 (*SC* 268. 178–80); on temptation bringing self-knowledge, *Or.* 29. 15, 17 (*GCS* 3. 390, 391–2).

[121] *Or.* 22. 1–4 (*GCS* 3. 346–9); *Jo.* 20. 34 (*GCS* 10. 306); *Com. Eph.* Fr. 3 (*JTS* 3 (1901–2), 237).

'brother' of Christ, and finally a 'son' of God.[122] Unity with God is therefore expressed in the continual activity of the soul in relation to God, and is indeed related to the union between the Father and the Son.

The task of the creature is therefore to imitate the perfect union of will and nature in God which is the unhindered expression of the good; what is God's by nature, humans acquire by will. Humans must choose their proper being as children of God—that is, recognize their source of existence and dwell continually in relation to God. Those who refuse slide down gradually toward the abyss of non-being. Equally important, 'sonship' consists not just in an intellectual recognition of origins, but in continuous active love in imitation of God in the living of everyday life. This is possible because of the activity of God throughout creation, particularly in the perfect response of the Son and the faithfulness seen in the life of Jesus. As God is good and acts justly, so a person must act justly to become good. He or she is created to be good but must choose to continue in goodness and being. Transformation therefore entails a stability of character in continual exercise, rather than a motionless perfection.

Origen's explanation of Christian life through a blend of Stoic psychology and Platonic metaphysics meant that being was intimately related to moral actions. He did not confuse nature with the supernatural, or transcend morality in mystical knowledge, but constructed a universe whose very motion and structure lay in the ongoing tension between God's grace and human freedom. The gradual decline into evil, the struggle for goodness, and the force of habit on evil or good are all descriptions of individual moral struggle, rather than analytical accounts of human free will. Lacking the clear historical structure of Irenaeus's theodicy, Origen's account can be, and has been, read as an account of cosmic and disembodied cycles of virtue in which the contrast between static perfection and embodied changeability is never overcome.[123] On the other hand, by contrast with ancient fatalism, Origen's uncompromising, broad defence of divine will and

[122] On progress to knowledge through perseverance, see *Princ.* 2. 11. 3 (*SC* 252. 398–400); *Cel.* 1. 57 (*SC* 132. 230); *Hom. Jer.* 9. 4 (*SC* 232. 392); *Jo.* 20. 33 (*GCS* 10. 370); cf. Irenaeus, *Haer.* 4. 24. 3.

[123] Otis, 'Cappadocian Thought as a Coherent System', *Dumbarton Oaks Papers*, 12 (1958), 95–124.

human freedom communicates much of the spiritual liberation and daily discipline experienced by ancient Christians. To be oneself is to be in free communion with God, as depicted most profoundly in the teachings and example of Christ.

Christology

Origen's attempt to explain the incarnation of the Logos in terms of a pre-existent human soul was one of the first constructive Christologies, and anticipated many problems in later theology regarding the proper union of divine and human nature. By examining the patterns of agency and obedience of Christ outlined in his exegesis, theological constructions, and preaching, we come to a clearer understanding of Origen's conception of the incarnate Logos as exemplar of the relationship between divine and human will. Ultimately, Origen's optimistic and dynamic cosmology pivoted on the divine will and human freedom as revealed in the actions of Christ.

As discussed above, Origen affirmed a unity of essence between Father, Son, and Holy Spirit, based on their eternal relations to one another. The Father and the Son have a common *οὐσια*, in the sense of a shared divinity, which does not infringe on the uniqueness of the Father or leave the Son as merely one of the created. Even if Origen had used the term *ὁμοούσιος* of the Father and the Son, he would have understood it in a way which preserved the primacy and uniqueness of the Father. In fact, Origen's extant use of the term is confined to the Latin fragments, and seems to contradict his reactions to modalist and Gnostic discussions of substance language.[124] The Father and the Son share a certain substratum, yet are distinct according to *ὑποκείμενον, ὑπόστασις*, and *οὐσία*.[125] Wary of the language of emanation, Origen was ambiguous as to whether the Son was generated from the *οὐσία* of the Father.[126] In his discussions of divine substance, Origen does not deny the divinity of the Son; but he distinguishes carefully the persons of the Father and the Son. Thus, he may say that they share the qualities of goodness and divinity absolutely, in a manner distinct from creatures; yet

[124] Stead, *Divine Substance*, 209–14.

[125] *Jo*. 2. 75 (*SC* 120. 254), 6. 188 (*SC* 157. 268); *Or*. 15. 1 (*GCS* 3. 334).

[126] *Jo*. Fr. 109 (ed. Brooke, 2. 311); cf. 20. 18 (*SC* 290. 157–8).

he sees these qualities being possessed in a different manner by the Father and the Son. The essential characteristics of the persons in the Godhead are distinguished by causality and operation.

The means of causality are of course critical in any description of divine nature. Origen rejects any literal understanding of generation, but acknowledges an exceptional process worthy of God. Drawing on Wisdom 7: 26 and Hebrews 1: 3, he describes the Son in *On First Principles* as breath or brightness or the image of the Father's substance.[127] In his discussions of these scriptural metaphors, he emphasizes the intimacy of the Son as directly and eternally proceeding from the Father, yet not as an emanation or temporary economic separation, but as a distinct ὑπόστασις. The description of generation thus underlines the dynamic relation within the Godhead. The very energy and activity of the Son ensure his independence, whereas the will of the Father safeguards his eternity and existence:

Huius ergo totius vertutis tantae et tam inmensae vapor et, ut ita dixerim, vigor ipse in propria subsistentia effectus quamuis ex ipsa virtute velut voluntas ex mente procedit, tamen et ipsa voluntas dei nihilominus dei virtus efficitur. Efficitur ergo virtus altera in sua proprietate subsistens. (*Princ.* 1. 2. 9 (*SC* 252. 130))

In a sense the Son is an ἐνέργεια of the Father. Yet the very force of the divine nature results not in a channel of energy, but in a separate and active power in relation to the original.[128]

Origen's reference to the will (*voluntas*) of the Father in the generation of the Son is undoubtedly a safeguard against emanation or corporeal images of generation. In the early work against Candidus, Origen uses will to refute the Gnostic concept of προβολή.[129] In *On First Principles* he uses will in generation in two ways. The more common is the desire of God for Wisdom as the reason for his existence. This perpetual choosing of the Son reflects God's eternal will behind all existence:

Et ita semper de anterioribus inquirentes et verbo interrogationis ascendentes perveniemus in illum intellectum, ut quoniam semper et poterat

[127] *Princ.* 1. 2. 4–5 (*SC* 252. 112–16).

[128] Lorenz, *Arius judaizans?*, 93.

[129] *Con Ruf.* 2. 19, cited in Nautin, *Origène*, 169; cf. Crouzel, *Théologie de l'image de Dieu*, 90; A. Orbe, *Hacia la primera teologia de la procession del Verbo* (Rome, 1958), 399.

deus et volebat, numquam vel decuerit vel causa aliqua existere potuerit, ut non hoc, quod bonum volebat, semper habuerit. (*Princ.* 1. 2. 9 (*SC* 252. 41))

The Son as Word, or Wisdom, was therefore always with God as an effect of the eternal will of God; thus, Origen may say that the Son has no beginning because he began in the Father.[130] As discussed above, the sense is one of continuous choosing rather than eternal outflow. Given Origen's conception of the coincidence of divine will and nature in God, the Son is always begotten. In a later homily Origen compares the continuous action of good works which makes humans adopted 'sons' with the Father's action which eternally begets the Son.[131]

Closely associated with this is Origen's analogy of the Son's generation as will from the mind. This image occurs three times in *On First Principles* and once in the *Commentary on John*.[132] In each case Origen explicitly rejects any notion of emanation or physical process by referring to the process of will as an analogy. Thus he does not identify the Son literally as the will of the Father; the will of the Father remains separate as the cause of the Son or the analogy closest to the process of generation. Thus, the straightforward sense of Origen's description is the Father's desire and power as the origin of the Son:

Si enim 'omnia quae facit pater, haec et filius facit similiter', in eo quod omnia ita facit filius sicut pater, imago patris deformatur in filio, qui utique natus ex eo est velut quaedam voluntas eius ex mente procedens. Et ideo ego arbitror quod sufficere debeat voluntas patris ad subsistendum hoc, quod vult pater. Volens enim non alia via utitur, nisi quae consilio voluntas profertur. Ita ergo et filii ab eo subsistentia generatur. (*Princ.* 1. 2. 6 (*SC* 252. 122))

As discussed above, divine nature and will coincide, so that Origen's use of will in generation does not necessarily imply a created status for the Son or his existence as a mere channel of energy. In each case he underlines their separate, if continuous existence, by the power of the Father's will to generate the Son

130 *Princ.* 1. 2. 9 (*SC* 252. 130). See M. F. Wiles, 'Eternal Generation', *JTS*, NS 13 (1961), 284–91.

131 *Hom. Jer.* 9. 4 (*SC* 232. 392–4).

132 *Princ.* 1. 2. 6 (*SC* 252. 122), 1. 2. 9 (*SC* 252. 130), 4. 4. 1 (*SC* 268. 399); *Jo. Fr.* 108 (ed. Brooke, 2. 310–11).

as a separate existence and by the activity of the Son as part of his being the image of God.

The Son therefore possesses divinity through his direct derivation from the Father by will and nature. Yet the primacy of the Father as ungenerated in Origen's cosmology puts all other existence, including the Son, into contingent relation with God; hence the Son, in spite of his eternal generation, can be called both God and creature.[133] The Son is divine through being first-born, eternally begotten from God, and participates directly in the Father. He is of a derivative, but unquestionable divinity. Equally important, the Son's activity within creation separates him from the Father, yet also reveals the Father in the manner of a Middle Platonic second principle. In his acts of love the Son reveals the divine essence, for he is the 'image of the invisible God'.[134] Yet, by his very activity in incarnation and death, the Son is of a different substance from the Father. The mediating character of the Son's essence is linked not only to his generation by will from the Father, but also to his role as the active agent within the cosmos as Wisdom, Word, and Saviour. In Stoicism this is the process of individuation within shared being; the different activity produces a distinct, if essentially related individual.[135]

The Son reveals facets of divine nature through his complex of names, the *ἐπίνοιαι*. As studied extensively by Harl and Gruber, Origen ordered the titles of the Son so as to correspond to the hierarchy of being and the spiritual growth of the Christian.[136] As one grows in faith, one is able to move beyond physical appearances and moral discipline to full participation in the Son and, through him, the Father. A common suspicion of this doctrine has been that the life and death of Jesus constitute the lowest stage in the accommodation of the Logos, and that history and material life are left behind as one penetrates the deeper, spiritual mysteries of God.

Ironically, the complexity of Origen's Christology lies not

[133] Divinity by proximity in *Jo.* 2. 18 (*SC* 120. 218). Cf. C. W. Lowry, 'Did Origen Style the Son a *ΚΤΙΣΜΑ*?', *JTS* 39 (1938), 39–42.

[134] *Jo.* 1. 230–2 (*SC* 120. 172–4), 6. 295 (*SC* 157. 352–4), 13. 234 (*SC* 222. 156–8).

[135] Voelke, *L'Idée de volonté*, 13–18.

[136] *Jo.* 1. 119f. (*SC* 120. 122f.), 19. 37 (*SC* 290. 68). On categories, see G. Gruber, *ΖΩΗ* 104f.

merely in the fact that Christian life is explained on the basis of a Platonic grid of hierarchical being, but that these levels of being are linked directly with the life of Christ as reported in Scripture. As in contemporary Platonism, the *αυτο-* prefixes describe the Son as the bridge between the noetic and sensible worlds; yet many of the titles are themselves drawn from Christ's soteriological activities as delineated in Scripture.[137] Thus Origen links Mark 10: 18 and Wisdom 7: 26 to contrast the unchanging goodness of the Father with the secondary, active goodness of the Son.[138] However, the activities which define the Son over against the Father are not only those of cosmological or providential agency, as in contemporary Platonism, but those of the incarnate Jesus as revealing the Father or eating with sinners.[139] Although the Father and the Son are both called *φῶς*, the Son is distinct because he shines in the darkness, as a result of the incarnation; the Son may not be called immortal by nature, as is the Father, because he once died.[140] Not only the Platonic ontological hierarchy, but the specific biblical and soteriological roles of the Son thus shaped Origen's account of the secondary essence of the Logos. His belief in the profound union between human nature and divine nature accomplished in Jesus Christ, the incarnate Word, led Origen to blend these soteriological activities in the cosmological relation between the Father and the Son.

Against adoptionism and Gnosticism, he therefore underlines the union of the natures by attributing various activities to the one subject.[141] Yet he distinguishes some activities as appropriate to the humanity or the divinity of the incarnate Word for theological or exegetical clarity. For example, it is not the Logos that is exalted to God's right hand, but rather the Son of Man.[142] Scholars have noted the difficulty of Origen's Christology at this point, in that he seems indecisive in applying all titles to either the Logos or Jesus in a consistent pattern.[143] However inconsis-

137 *Jo.* 1. 9 (*SC* 120. 88–90).

138 *Princ.* 1. 2. 13 (*SC* 252. 140–2); *Com. Mt.* 15. 10 (*GCS* 40. 374–7). See H. Saffrey, 'Les Extraits du *περὶ τἀγαθοῦ* de Numenius dans le livre XI de la Préparation évangelique d'Eusèbe de Cesarée', *SP* 13 (1975), 50.

139 He cites Phil. 2: 6 and Wisd. 7: 26 in *Jo.* 1. 230–2 (*SC* 120. 173–4), and Mark 10: 18 in *Jo.* 6. 293–5 (*SC* 157. 352–4).

140 *Com. Rom.* 4. 10 (*PG* 14. 998–9); *Jo.* 2. 123–42 (*SC* 120. 288–300).

141 *Cel.* 2. 9 (*SC* 132. 300–6); *Princ.* 2. 6. 3 (*SC* 252. 316).

142 *Jo.* 32. 25 (17) (ed. Brooke, 2. 198).

143 Harl, *Origène et la fonction révélatrice*, 358; cf. 198–224.

tent judged by later patterns of Christological exegesis, Origen's determination to do justice to both Logos theology and the descriptions of Jesus in the New Testament reveal the synthesis of activity and essence in his cosmology. The incarnate, mediating Logos can as one subject represent both divine accommodation and human participation:

> One should pray to the supreme God alone, and pray also to the only begotten and first-born of all creation Word of God; and one ought to ask him as a high priest to bear our prayer when it has reached him, up to his God and our God and to his Father and the Father of those who live according to the word of God. (*Cel.* 8. 26 (*SC* 150. 232))

This passage mixes cosmological mediation—that is, access to the supreme God through the Logos—with attention to Christ's role as the great high priest who has passed through the heavens in Hebrews and the risen Christ of John 20: 17. Soteriology and cosmology, historical process and spiritual ascent, are united in the activity of the one subject.

The human life of Jesus not only shapes the cosmological role of the Logos in Origen's Christology, but is the model and exemplar for Christian devotion. Origen's descriptions of the human soul assumed by the Logos, even in discussions of the Christological union, reveal the depth of his fundamental concern for human freedom and progress towards unity with God. In the early *Commentary on John* Origen refers to the historical example of Jesus as the lowest, somatic level of the gospel; while necessary for the simple believer, it should be passed through quickly.[144] Unfortunately, many commentators have tended to follow his advice, focusing on the actions of the Logos. Thus, to avoid a troublesome separation of this soul in 'Nestorian' fashion, Crouzel suggested that the soul should be seen as the metaphysical principle which unites divinity and humanity.[145] Yet the full context of Origen's discussion of the soul suggests a far more important role. Like the souls which fall away, this soul was essentially *τρέπτος*, yet remained with the Logos by *προαίρεσις*.

[144] *Jo.* 1. 51 f. (*SC* 120. 88–90); cf. R. Greer, *The Captain of Our Salvation* (Tübingen, 1973), 52 f.

[145] *Princ.* 2. 6. 3 (*SC* 252. 314); Crouzel, *Théologie de l'image de Dieu*, 242–3. See also Grillmeier, *Christ in the Christian Tradition*, 146.

Since the faculty to choose good and evil is within all, this soul which is Christ's chose to love [*diligere*] righteousness and by its great love cling unchangeably and inseparably, so that by firm determination, great desire, and inextinguishable heat of love [*propositi firmitas et affectus inmensitas et dilectionis inextinguibilis calor*] all inclination to change or alteration was destroyed, and what had depended on will by prolonged exercise of desire changed into nature [*quod in arbitrio erat positium, longi usus affectu iam versum sit in naturam*]. (*Princ.* 2. 6. 5 (*SC* 252. 318–19))

The theological context is Origen's dynamic ontology based on the transformation of created souls by προαίρεσις in relation to divine power. Christ's soul is therefore the archetype for the reunion of all souls with God through the Logos; action and essential being are not confused, as Grillmeier suggests. Rather, their unity reflects the determination of the quality of being by action in Origen's ontology.[146] Origen's language about the Christological union is at times deliberately active and quantitative, so as to preserve the true humanity of the soul: 'supreme participation' or 'full anointing'.[147] It is therefore not unnatural for a soul to receive divine qualities, since all souls are by nature related to the Logos; but Christ's soul alone preserved the true image of the image of God.[148]

By the phrase 'one nature' Origen may be hinting at the fusion of wills which, through longevity, becomes nature in the sense of habit, rather than a metaphysical transformation as in later Monophysite theology: if one were to become one spirit with God (1 Cor. 6: 17), this soul could become one nature with the Logos.[149] The union is not therefore 'merely moral', in the sense of an external co-ordination of wills, since for Origen the union of action brings about ontological change in the sense of qualities of being; yet it is not wholly metaphysical either, since the distinction of human and divine remains. In his participatory cosmology the human soul remains free, but its total congruence

[146] Grillmeier, *Christ in the Christian Tradition*, 146–7.

[147] *Princ.* 2. 6. 6 (*SC* 252. 320–2); *Cel.* 5. 39 (*SC* 147. 118–20), 6. 47 (*SC* 147. 298); cf. *Cel.* 3. 41 (*SC* 136. 96).

[148] *Princ.* 2. 6. 5 (*SC* 252. 318–20); *Cel.* 3. 41 (*SC* 136. 96).

[149] *Princ.* 2. 6. 3 (*SC* 252. 316); *Dial.* (ed. Scherer, 126); *Jo.* 32. 25 (*GCS* 10. 470).

with the Logos makes it 'divine': the soul is human, but the will is from the Logos.[150]

In his discussions of the Christological union, Origen is careful to preserve the free action of Christ's soul as part of its essential humanness; but in his homilies he discusses the perfect obedience of Jesus as the key image of human freedom and spirituality for ancient Christians. He uses Psalm 45: 7 and Isaiah 7: 15–16 to describe the pre-existent virtue of Jesus which God foresaw, but did not determine.[151] This is one of Origen's strongest statements against predestination and contemporary fatalism. Because of his voluntary virtue, Jesus may be considered to be the height of goodness, just as the devil, by voluntary evil, conveys the depth of sin; by this virtue and union with God, he is halfway between the uncreated and the created.[152] While on the one hand Origen clearly affirms the unchanging and unique union of Christ's soul with the Logos, he also refers to Jesus as our brother, the pioneer who 'progressed' towards God, citing Luke 2: 52 and Hebrews 2: 10.[153] This appeal to the example of Jesus as 'progressing' seems to refer to his continual, perfect harmony with the Logos throughout his life, rather than to any actual change as a result of virtue. For Origen, Jesus was the exemplar by his achieved perfection: 'He flourished in the vigour of his soul, his spirit made great by his works.'[154] His continuous moral perfection rooted in his union with God is thus the model for Christian imitation:

> We have a great high priest . . . Jesus Christ, who never leaves the holy of holies, but always is in the holy place, always holy in his words, holy in his acts, holy in all his desire and will [*sanctus in omnibus voluntatibus suis*], and therefore never leaves. All who sin leave the holy of holies. But Christ who did not sin, never leaves. But you, who follow Christ and imitate him, if you remain in the Word of God, meditating on the law day and night, you also will remain in the holy of holies. (*Hom. Lev.* 12. 4 (*SC* 287. 180–2))

[150] Crouzel rejects the moral sense, but underestimates the importance of προαίρεσις (*Théologie de l'image de Dieu*, 242).

[151] *Princ.* 2. 6. 4 (*SC* 252. 316), 4. 4. 4 (*GCS* 268. 354–5).

[152] *Cel.* 6. 45 (*SC* 147. 292), 3. 34 (*SC* 136. 80).

[153] *Hom Lev.* 12. 3f. (*SC* 287. 168); *Hom Jer.* 14. 10 (*SC* 238. 87: we advance as Jesus did in the Logos); *Com. Mt.* 13. 26 (*GCS* 40. 250–1).

[154] *Hom. Lev.* 12. 2 (*SC* 287. 168).

Stability through obedience and love are the goals for Christians to imitate in everyday spiritual life. When one is tempted, one should remember Jesus who was tempted for our sake.[155] He is our brother, advocate, and leader in prayer to the Father.[156] The perfect virtue of Jesus is rooted in his unchanging union of will with the Logos, offering a model for individual Christians.

In the case of the exegetical subject being the Son or Logos, the sense of will was not one of autonomous action, but rather of agency in accordance with God's active good will in creation. The language about the Logos was often closer to that of a Stoic force of nature rather than a personal sense of movement or obedience.[157] Yet, when the language was applied to the 'Father' and the 'Son', it was coloured by scriptural references and took on more personal and volitional tones:

> And so the only begotten Son of God, Word and Wisdom of the Father . . . 'emptied himself and taking the form of a servant became obedient even to death' to teach obedience to those who could in no other way obtain salvation except through obedience. . . . He first fulfilled in himself what he wished to be fulfilled in others. (*Princ.* 3. 5. 6 (*SC* 268. 228))

Here, between the Father and the Son, as in the discussion of the union of Jesus with the Logos, is a sense of active conformity between two wills rather than an automatic congruence of nature. The language of obedience in reference to a cosmological principle makes sense only in terms of Christian theology in which the incarnate Son is the pivot between the will of the Father and a model for human action. Thus, Origen adds *ἑκούσιος* to Philippians 2: 6 to underline the voluntary submission of the Word to becoming incarnate.[158] On a cosmic level, this language refers of course to the Son as the active image of divine love, rather than to any decision; yet the scriptural presentation of the obedience of Jesus also informs the descent of the Word. Origen thus applies various levels of language of will

[155] *Luc. Hom.* 29. 6 (*SC* 87. 366).

[156] On Jesus as an example in prayer, see *Or.* 13. 1 (*GCS* 3. 325f.), 15. 1–4 (*GCS* 3. 333–6). On Jesus who leads our prayers to God, see *Or.* 16. 1 (*GCS* 3. 336); *Cel.* 5. 4 (*SC* 147. 21–2), 8. 26 (*SC* 150. 232), On Jesus as our brother, see *Jo.* 32. 30 (19) (ed. Brooke, 2. 206); *Com. Mt.* Fr. 243 (*GCS* 41. 113); *Hom. Lev.* 12. 2 (*SC* 287. 172).

[157] *Cel.* 8. 72 (*SC* 150. 340), 3. 68 (*SC* 136. 156).

[158] *Cel.* 4. 18 (*SC* 226); cf. 6. 15 (*SC* 147. 216).

within the natures of the Christology, with 'Son' bearing nuances of both agency and at times voluntary action. If, philosophically, the Logos is the active agent of the Father's will, the religious significance of the Incarnation, together with the personal language of Scripture, often cast the activity of the Word in more voluntarist terms.

The relation between divine nature and will in Origen's Christology may be further clarified by examining his use of the term *χωρέω*. In the *Commentary on John* the union of the will between the Father and the Son is described as follows: '*καὶ γενέσθαι τὸ θέλημα τοῦ υἱοῦ ἀπαράλλακτον τοῦ θελήματος τοῦ πατρὸς, εἰς τὸ μηκέτι εἶναι δύο θελήματα ἀλλὰ ἓν θέλημα· ὅπερ ἓν θέλημα αἴτιον ἦν τοῦ λέγειν τὸν υἱὸν ἐγὼ καὶ ὁ πατὴρ ἕν ἐσμεν.*'[159] Because *θέλημα* conveys the immutable nature of God, the Son's will which is congruent with the Father's thus expresses their shared divinity: '*καὶ διὰ τοῦτο τὸ θέλημα ὁ ἰδὼν αὐτὸν ἑώρακε τὸν υἱὸν, ἑώρακε δὲ καὶ τὸν πέμψαντα αὐτόν.*' Hence the Son by his actions contains the Father as his image: '*ἅπερ βούλεται τὸ θέλημα τοῦ θεοῦ· μόνος δὲ ὁ υἱὸς πᾶν τὸ θέλημα ποιεῖν χωρήσει τοῦ πατρός· διόπερ καὶ εἰκὼν αὐτοῦ.*' The mention of image, together with John 10: 30 and 12: 45, links this passage with *On First Principles*, 1. 2–8, where the Son is described as the image of the Father as a result of his actions. But the Son is again both the revealer of divine will and the model of human response, since all may contain the will of God, like Christ, and become one with God.[160] The effect of the perfect will of the Son is to create a similar effect in creatures, thereby making them adopted 'sons' of God. To contain (*χωρέω*) the will of God implies a conformity of purpose which allows a share in the life of God, by receiving love, knowledge, and goodness.[161]

Origen's descriptions of the will of the Son may be criticized as conveying neither authentic human struggle nor, within his hierarchical ontology, the will of the Father.[162] As discussed above, Origen affirms that the locus of salvation lies in divine power,

[159] *Jo.* 13. 36 (*SC* 222. 156–8). [160] *Or.* 26. 3 (*GCS* 3. 360–1).

[161] On the progress of the individual sharing in the life of God, see *Jo.* 32. 27 (ed. Brooke, 2. 201), 32. 30 (19) (ed. Brooke, 2. 205–6).

[162] R. Williams, 'Origen on the Soul of Jesus', in R. P. C. Hanson (ed.), *Origeniana Tertia* (Rome, 1985), 131–7.

not in the moral obedience of a human being: Christ came not with mere sympathy, but with power.[163] Jesus, therefore, is unique in his relation to his body and able to change it at will.[164] In Late Antiquity, however, Jesus's unchanging will does not ultimately undercut the Son as exemplar for the perfect heroic Christian. In Gethsemane Jesus was not afraid, but, like the enthusiastic martyrs in the arena, sought a more courageous death:

Since the Saviour through the illumination and salvation given to him by the Father fears nothing . . . secretly he asked for another (death) which would have been more of an ordeal, so that by a different cup he might achieve more benefits. This was not the will [θέλημα] of the Father which was wiser than the will [βούλημα] of his Son or the Saviour's vision as he ordered the economy of the events. (*Ex. Mart.* 29 (*GCS* 2. 26))

Like the stalwart martyrs in the accounts of the second and third centuries, the Son embodies a perfect, unchanging confidence in God's will. The issue for Origen in this conflict of wills is not the fear of the Son, for Origen denies it, commenting in relation to the martyrs, 'If he was afraid, who ever was courageous?' The problem for Christ, as for all, was the steady co-operation with the Father's will in the broader economy of salvation; whether in the pre-existent or the incarnate state, the Father's will is the guiding force for the activity of the Son, as for any Christian.[165] As Origen pointed out against Heracleon, this will, as described in Gethsemane, is not one of rest or deliverance, but of power (δύναμις).[166]

In Origen's exegesis and preaching, therefore, the nature and will of Christ are described on levels of being which embrace both humanity and divinity: the cosmic agency of the divine Logos in accomplishing and revealing the power of God, the self-

[163] *Hom. Jer.* 15. 6 (*SC* 238. 126): we do not hope in a man, but in wisdom.

[164] *Cel.* 2. 64–5 (*SC* 132. 434f.); *Jo.* 19. 16–18 (*SC* 290. 108–16). On the uniqueness of Jesus, see J. A. Alcain, *Cautiverio y redencion del hombre en Origenes* (Bilbao, 1973), 313–14, and M. Eichinger, *Die Verklärung Christi bei Origenes* (Vienna, 1969), 202.

[165] Ignorance of divine will can preserve human effort; see *Com. Gen.* 6. 11 = *PE* 6. 11 (*GCS* 43. 1. 354). On following Jesus in suffering, see *Ex. Mart.* 13 (*GCS* 2. 13), 37 (*GCS* 2. 34); *Jo.* 13. 242–6 (*SC* 222. 160–2). On the reality of temptation and Jesus as a great wrestler, see *Cel.* 1. 69 (*SC* 132. 270).

[166] *Jo.* 13. 249 (*SC* 222. 164).

emptying love and obedience of the Son, and the perfect human virtue of Jesus. These levels of action, like the levels of revelation, portray the varied accommodation of divine goodness in the interests of human salvation. In Origen's dynamic, anti-fatalistic cosmology, willed actions alter and define one's relation to God; and hence the work of the Logos is to ensure the transformation of individual humans towards God by both power and revelation. This transformation is seen most profoundly in the life of Jesus, who was the human exemplar and revealer of the divine Word. Against ancient fatalism, the actions of Jesus remain a touchstone for Christian endurance and salvation: 'As sons of a patient (*μακροθυμία*) God and brothers of patient Christ, let us be patient in all that may happen to us.'[167] The eternal Son loves (*προαίρεσις*) the Father, and is one with him in will and essence; he is eternally begotten. The human soul is united by virtue and love (*προαίρεσις*) to the Logos, and by a transformed will and by contemplation recovers what the Son enjoys by nature: unchanging union with God. The pattern of relationship between the Father and the Son, as well as that between the soul and the Logos, point towards the final restoration of all humanity with God.

These cosmological levels of the work of the incarnate Logos give a penetrating, optimistic focus to the purpose and goal of ancient Christian life. Against the Marcionites and certain Gnostics, Origen defended the immediate power of goodness throughout creation and the freedom of individuals to appropriate it. Material existence hinders, but does not preclude eventual divinization; bodies are capable of being transformed by divine will and individual human discipline. Because the congruence between nature and will in God assures salvation and the imbalance in human nature prevents it, both revelation and obedience are emphasized by Origen in his Logos Christology. As the pivot in his dynamic ontology, Christ in two natures exemplifies not only divine will, but also human obedience. This aspect of his Christology needs to be held together with the revealing, pedagogical work of the Logos. The religious goals exemplified by Christ's life are essential freedom from fate and demons, the possibility of a steady course of virtue through union with the

[167] *Ex. Mart.* 43 (*GCS* 2. 40).

Logos, and ultimately complete union with God.

Amidst the intellectual and social ferment of Alexandria, Origen affirmed the possibility of individual progress through the spiritual instruction of the Logos by defending divine goodness and power as reflected in the very structure of the cosmos. The Christological *ἐπίνοιαι* point to the diversity of the urban Christian community engaged in moral discipline and theological reflection. As a teacher, by both revelation and example, Christ is the powerful pivot between human struggle and divine purpose, communicating power and portraying the possibilities of human perfection. In spite of the ontological gulf between divinity and humanity, divine love, as a free and continual relationship, is able to unite each individual with God. Origen's cosmology thus reflects the varied experience of his community and his own personal engagement with the spiritual development of Christians, by emphasizing the slow transformation through discipline and instruction of various souls on their particular journey toward God.

3

EUSEBIUS OF CAESAREA: POWER AND PROGRESS

> This is the greatness of our Saviour's mission for us that we may be with him where he is, and so that we may see his glory for he loves us just as his Father loves him; for as he was given, he gives to us, and the glory he was given he gives to us; . . . For by his wisdom, judgment, righteousness, piety, and perfect virtue we are lit by the Father's unspeakable divinity . . . and become sons of God.
>
> Eusebius, *The Ecclesiastical Theology*

The most recurrent description of Eusebius's theology is 'old-fashioned'; it is usually summarized as a mixture of biblical exegesis, simplified Origenism, and uncritical Platonism.[1] The bishop of Caesarea was an industrious, skilful scholar whose encyclopaedic works preserve much precious information; yet few turn to him for theological insight. In the eyes of his contemporaries, however, Eusebius was not only a scholar and a bishop, but an active participant in theological controversies. He defended Origen with his teacher Pamphilus, chastised Bishop Alexander of Alexandria theologically for his treatment of Arius before Nicaea, pleased Constantine with his imperial interpretation of salvation history, and defended Asterius as a traditionalist against Marcellus of Ancyra.

[1] T. Berkhof, *Die Theologie des Eusebius von Caesarea* (Amsterdam, 1939), 37; F. Ricken, 'Die Logoslehre des Eusebios von Caesarea und der Mittelplatonismus', *Theologie und Philosophie*, 42 (1967), 314–58; idem, 'Nikaia'; idem, 'Rezeption'; R. Grant, *Eusebius as Church Historian* (Oxford, 1980), 165; T. D. Barnes, *Constantine and Eusebius* (Cambridge, 1981), 100; Young, *Nicaea to Chalcedon*, 17–18. Recent studies of Eusebius's theology continue to emphasize his conservatism and his Platonism; see Lorenz, *Arius judaizans?*, 204–10; C. Luibheid, *Eusebius of Caesarea and the Arian Crisis* (Dublin, 1981); Williams, *Arius*, 174; Hanson, *Christian Search*, 48–58.

For better or worse, Eusebius's theological ambiguities reflect much of early fourth-century Christianity. Recent studies have emphasized the contemporary plurality of perspectives and tools in theological reflection; if a Logos Christology had become *de rigueur*, it did not carry many specific definitions for understanding the relation between God and humanity.[2] In the view of many scholars, Eusebius's Logos theology places him in the category of cosmological subordinationists: his uncritical enthusiasm for Platonism shapes his account in the direction of cosmological mediation rather than soteriology.[3] If comparisons with Origen or Nicene orthodoxy have usually found Eusebius wanting, a reconstruction of his Christology in the light of the shifting cosmological and soteriological categories of the fourth century may reveal his particular strengths in defending divine power and human progress. From his earliest apologies to his final summation, *The Theophany*, Eusebius celebrated the miraculous will of God in history, which, through the teaching and incarnation of the Logos, altered human society and life for ever.

EUSEBIUS AS APOLOGIST

From the death of Origen in 254 until the Council of Nicaea in 325, Christians lived through periods of relative peace to a final bitter persecution under Diocletian and the heady legitimacy under Constantine. This political security and social self-definition strengthened their unity, but led eventually to a harsher response to theological controversy, as Christians began to possess a clearer sense of their beliefs and a means to enforce them.[4] Born around 270 and elected bishop in 314, Eusebius reflected many of these issues of identity and unity in his life, writings, and ecclesiastical roles. The apologetic, optimistic

[2] On the theological transitions of the late third century, see R. Sample, 'The Christology of the Council of Antioch (268 C.E.) Reconsidered', *CH* 48 (1979), 18–26; Williams, *Arius*, 149–74.

[3] H. G. Opitz, 'Euseb von Caesarea als Theologe', *ZNW* 34 (1935), 17–18; Berkhof, *Theologie*, 30; A. Dempf, 'Der Platonismus des Eusebius, Victorinus und pseudo-Dionysius', in *Sitzungsberichte* (1962), no. 3, 7; Ricken, 'Rezeption', 348; idem, 'Nikaia', 341; Luibheid, *Eusebius and Arian Crisis*, 122; Lienhart, '"Arian" Controversy', 424; Hanson, *Christian Search*, 48.

[4] R. A. Markus, 'The Problem of Self-Definition', in Sanders (ed.), *Jewish and Christian Self-Definition*, i. 1–15.

theology of Justin and Origen was foundational to his work, whether his reading of philosophy, his defence of the history of the Church, or his descriptions of the deaths of martyrs. Eusebius was not a speculative thinker, but a man of texts and tradition. As a Christian historian, he defended the efficacy of human decision and the possibility of discerning the pattern of divine will in past events. If Greek historians wrote to instruct society, Eusebius wrote to instruct both Christians and, hopefully, Graeco-Romans on the divine economy unfolding in the history of the world.[5] If his history was therefore essentially apologetical, his theology was rooted in his extensive study and belief in the reality of salvation history. The life and beliefs of the Church must be justified broadly and openly in order to vindicate Christian identity.

Like those of many of his predecessors, Eusebius's theological and apologetic interests took form against a background of internal Christian controversy and Roman persecution. Born in Caesarea, Eusebius became a protégé of Pamphilus, and inherited the riches of Origen's library, as well as a commitment to apologetics. His earliest extant work, *Against Hierocles*, written in 303, was a treatise directed against Roman criticism of Jesus's uniqueness. His historical and scriptural interests are also seen in his *Eclogae Propheticae*, which traces the divine origin of Christ in prophetic literature. In these works the truth of Christianity as the fulfilment of history and the uniqueness of Jesus are defended by literary proof and rhetorical argument. During the savage persecution of Diocletian, Eusebius wrote an eyewitness account of the deaths of martyrs in Palestine and Egypt. To him, their painful deaths and determined courage testified to both a slack, divided Church and the redemptive power of God.[6]

With Pamphilus Eusebius composed a *Defence of Origen*. Written partly while they were imprisoned in 309, this work attempted to sort out the criticisms of Origen into those based on mere hypothesis, those stemming solely from hostility, and those which were unfortunately justified but not theologically serious.[7] In general, Origen's speculations on eternal creation or pre-

[5] G. F. Chesnut, *The First Christian Histories: Eusebius, Socrates, Sozomen, Theodoret and Evagrius* (2nd edn. Macon, Ga., 1986), 65 f.

[6] *HE* 8. 1–4; cf. Barnes, *Constantine and Eusebius*, 154f.

[7] *Apol. Orig.* (*PG* 17. 541–616); cf. Nautin, *Origène*, 110.

existence were not to be tolerated by the majority of Christian teachers. Moreover, Porphyry's work criticizing Christianity raised embarrassing questions about Origen's allegorical exegesis and Logos theology.[8] These criticisms precluded any continuation of Origenism as a coherent system, especially as regards Christology. Repudiated as implying two Sons, and probably suspect as adoptionist, Origen's Christology was replaced by a Logos–sarx model of incarnation in Eusebius and most of his contemporaries.[9] However, Eusebius remained one of the most faithful Origenists of his generation, although with critical modifications of Origen's theology.

After the martyrdom of Pamphilus and further imprisonment in Egypt, Eusebius, eyewitness of bloody persecution and guardian of theological tradition, was elected bishop of Caesarea, in 314. In the new era of toleration and the rising power of Constantine, Eusebius completed his series of historical and textual apologies, probably against the scathing anti-Christian legacy of Porphyry: *The Demonstration of the Gospel*, *The Preparation of the Gospel*, and the *Ecclesiastical History*.[10] These works celebrate the triumph of Christianity as the fulfilment of Jewish Scripture, Greek philosophy, and human history: 'Therefore, Christianity would be neither a form of Hellenism nor of Judaism, but something between these two: the most ancient institution for piety and the most venerable philosophy which is only recently codified for all people in the whole world.'[11]

Viewed in light of these scholarly and ecclesiastical interests, Eusebius's reaction to the Nicene controversy may become clearer. In support of Arius's defence of the Father's primacy as

[8] H. Crouzel, 'Les Critiques addressés par Méthode et ses contemporains à la doctrine origénienne du corps ressuscité', *Gregorianum*, 53 (1972), 679–716; L. Patterson, 'Methodius, Origen, and the Arian Dispute', *SP* 17.2 (1982), 912–23. On Porphyry, see A. Meredith, 'Porphyry and Julian against the Christians', *ANRW* II. 23. 2 (1980), 1119–49; T. D. Barnes, 'Porphyry *Against the Christians*: Date and the Attribution of Fragments', *JTS*, NS 24 (1973), 424–42.

[9] H. Riedmatten, *Les Actes du procès de Paul de Samosate* (Fribourg, 1952), 61 f.; A. LeBoulluec, 'Controverses au sujet de la doctrine d'Origène sur l'âme du Christ', in L. Lies (ed.), *Origeniana Quarta*, 223–37.

[10] The dating of Eusebius's works, especially the *Ecclesiastical History*, remains somewhat controversial, because of his tendency to revise and incorporate earlier materials in later works. Recent positions are summarized in Hanson, *Christian Search*, 47 n. 94, and Chesnut, *First Christian Histories*, 111–18.

[11] *DE* 1. 2 (*GCS* 23. 8).

an obvious, traditional teaching from Scripture and philosophy, Eusebius wrote to the bishop of Alexandria; for this he was condemned at the Council of Antioch in 324. At Nicaea he continued to defend the primacy of the Father; but he also affirmed the unique divinity of the Son, in line with traditional hierarchical Logos theology. Not surprisingly, he was uncomfortable with both radical Nicenes and radical Arians; defending his own Christology along with Asterius, he wrote against Marcellus in defence of traditional catechesis in the *Ecclesiastical Theology*. Recent evaluations of Eusebius portray him as traditional, rather than Arian, in his theological views.[12]

As a theological moderate and learned apologist, Eusebius was, not surprisingly, favoured by Constantine. However, although he has often been portrayed as a court theologian, T. D. Barnes and H. Drake both argue that this relationship was fairly formal and his visits to the court infrequent.[13] Constantine's patronage of Christianity confirmed Eusebius's historical and theological vision of the triumph of Christianity as the true universal religion. As an apologist rather than an imperial toady, Eusebius was happy to praise the new order as reflecting the fulfilment of divine will and Hellenic ideals. In revising his apologetic works to present the final summary of the triumph of Christianity, *The Theophany*, he retained his weathered optimism on the growth of Christianity throughout the Roman world under the cosmic and historical guidance of the Logos. He died in 339 or 340 as author of forty-six works and bishop for twenty-six years.

Literary attempts to defend Christianity by means of philosophical parallels with Scripture and historical outlines were not new, but Eusebius's efforts in this genre were self-consciously expansive. He knew and intended to surpass all apologies written before him. His logical proof of Christianity consisted in showing it to be the historical fulfilment of prophecy.[14] This historical

[12] Lorenz, *Arius judaizans?*, 210; Gregg and Groh, *Early Arianism*, 127 n. 136; Luibheid, *Eusebius and Arian Crisis*, 122f.; G. C. Stead, '"Eusebius" and the Council of Nicaea', *JTS*, NS 24 (1973), 85–100; Hanson, *Christian Search*, 48–58.

[13] Barnes, *Constantine and Eusebius*, 266; H. A. Drake, *In Praise of Constantine* (Berkeley, 1976), 8.

[14] In *HE* I. I. 3 Eusebius compares himself with his predecessors; cf. *DE* I. I. (*GCS* 23. 6). On his motivations, see also Grant, *Eusebius as Church Historian*, 39–40.

method was both traditional and suited to his temperament: Eusebius was a man of texts and experience, not of ideas. In his preface to *The Preparation of the Gospel*, he noted not only rhetorically:

> Our purpose is to be worked out in our own way. First, to depreciate deceitful and sophistical plausibilities and to use proofs free from ambiguity . . . But, all words are superfluous when works are more obvious and plainer than words, which the divine and heavenly power of our Saviour shows distinctly. (*PE* 1. 3 (*GCS* 43. 1. 10–11))[15]

Thus, for Eusebius, doctrine alone had little persuasive power; it was the evidence of the life of Jesus and the continuing power of the church that were ultimately convincing. In part, of course, this was traditional apologetics. However, Porphyry's attack on Christianity, based on a thorough knowledge of Hebrew prophecy and the New Testament, had sharpened critical perceptions of Christianity; fourth-century apologetics thus had to be based more extensively on scriptural and historical proof, as well as on doctrinal cogency.[16] Equally important, after the social crises of the late third century, religion in general focused on the problem of transcendence, particularly the shape and locus of divine power.[17]

In the extensiveness and repetition of Eusebius's apologetic and theological writings, one encounters the convictions of a tested churchman who believed in witness and events perhaps more than reflection: humans were convinced by acts, and then came to understand the meaning behind them; God was revealed by the character of the people who live in divine power.[18] Appealing to personal experience, as well as the evidence of history, literature, and Scripture, Eusebius anchored his definitive

[15] Cf. *PE* 1. 1. (*GCS* 43. 1. 18–19), 6. 6. (*GCS* 43. 1. 311–12), and quotation below, p. 90.

[16] Barnes, *Constantine and Eusebius*, 178f.; M. Smith, 'A Hidden Use of Porphyry's *History of Philosophy* in Eusebius's *Praeparatio Evangelica*', *JTS*, NS 39 (1988), 500–2.

[17] Brown, *Making of Late Antiquity*, 98. As Aristides commented on the power of Serapis: 'If we have said what he can do and what he gives, we have found who he is and what nature he has' (*Or*. 8. 1. 8, quoted in A. D. Nock, 'Studies in the Graeco-Roman Beliefs of the Empire', in Z. Stewart (ed.), *Arthur Darby Nock*, vol. 1 (Cambridge, Mass., 1972), 35).

[18] *DE* 3. 7 (*GCS* 23. 144), 3. 5 (*GCS* 23. 118f.), 3. 6 (*GCS* 23. 135); *HE* 10. 1. 8.

defence of the superiority of Christianity in the public discourse and known events of ancient Mediterranean life. Christian inevitability was thus defined and revealed most fully in the public arena. The transforming power of Christ was operative not only at the individual, but also at the communal and social levels, showing it to be the heir to true civic virtue and religion:

> So that now, as never before, the most exalted emperors . . . spit in the faces of dead idols and laugh at the ancient rites of their fathers . . . and confess Christ the child of God to be ruler of all and calling him saviour on monuments, write imperishably his righteous acts and victories in imperial characters in the midst of the city which is great among the cities of the world. (*HE* 10. 4. 16)

Theologically, Eusebius organized his historical and literary defence around a fundamental conviction of God's power as active and discernible at all levels of reality. Human experience, whether historical or spiritual, was grounded in the saving will and providence of God. Creation and redemption were part of God's continuing care through the Logos, recently incarnate in Jesus. This affirmation of God's freedom to create and redeem was of course commonplace among earlier Christian apologists such as Hippolytus or Irenaeus. Freedom of moral choice led to the creation of a theology of historical and cosmological drama as humans were lured back to their divine origin. As in Origen, cosmology and history were not separate, but were parts of the whole divine activity of individual redemption. In Eusebius, however, this activity became primarily public and social, revealing where and how human will and divine purpose interacted decisively; historical events were the primary means of communication and conversion.[19]

To present and prove this vision of reality, Eusebius drew on the wealth of intellectual riches available at Caesarea. The library begun by Origen and augmented by Alexander of Jerusalem and Pamphilus contained Christian sources, literary and historical works, apologetics, and philosophy.[20] Eusebius was perhaps atypical in Christian circles in his extensive reading of philosophical sources. Numenius and Atticus are preserved almost exclu-

[19] See Chesnut, *First Christian Histories*, 65–95.

[20] Grant, *Eusebius as Church Historian*, 41; des Places, *Eusèbe de Césarée commentateur* (Paris, 1982).

sively in his works, and he knew a great deal of Porphyry. More unusual was his knowledge of Plotinus, who in all probability was relatively unknown outside his own circle in the early fourth century.[21] Unfortunately, Eusebius did not discuss these materials, but generally marshalled quotations from them as proof texts for the antiquity and universality of Christian beliefs. Assuming the superiority of Hebrew wisdom over Greek thought, he often cited superficial verbal parallels between Scripture and philosophy.[22] These metaphysical *rapprochements* rarely appear profound because he did not develop or critique them; in fact, they may be longer versions of proof texts first assembled by Origen.[23] Often, an argument consists of a mosaic of texts which may convince by their bulk, if not their rational cogency.

If Eusebius tended to rely on volume of text rather than intellectual skill, he was a tireless and conscientious editor. He was one of the first ancient historians to cite sources verbatim so extensively; this he did precisely because human actions and words revealed the divine and universal plan.[24] Not an ecclesiastical antiquarian, but a shrewd apologist, Eusebius also continually revised his accounts in order to present Christianity in a light appropriate to current issues.[25] His creativity was thus revealed in his editing, his construction of parallels, and his determination to defend Christian identity as the fulfilment of both ancient Greek and Hebrew wisdom. He deliberately intermingled philosophical and scriptural categories to express the ancient universality as well as the novel particularity of Christianity.

Ironically, Eusebius has been accused of both historical and ontological bias in his work; he has been said to write neither accurate history nor orthodox theology. The theological charge is based on a perception of his work as Platonic and old-fashioned. However, this evaluation rests in part on a misunderstanding of the third-century issues which shaped his thought and in part on the standards of later orthodoxy. Not by nature a theologian, Eusebius nevertheless faced, rather than evaded, issues in

[21] J. M. Rist, 'Basil's "Neoplatonism"', in P. Fedwick (ed.), *Basil of Caesarea*, vol. I (Toronto, 1981), 159f.

[22] Cf. Numenius with John 5: 19 in *PE* 11. 18 (*GCS* 43. 2. 40–4).

[23] Saffrey, 'Les Extraits du περὶ τἀγαθοῦ', 80.

[24] Chesnut, *First Christian Histories*, 254.

[25] Grant, *Eusebius as Church Historian*, 164–5; Chesnut, *First Christian Histories*, 111–18.

contemporary theology, as shown in his treatises against Porphyry and his active role in the Nicene controversy. As a traditional apologist, he accepted an affinity between Christianity and philosophy; but, as will be shown, he did not avail himself of it uncritically, nor did he ignore the flaws of the church.

For Eusebius Christianity was neither Hellenic nor Judaic, but rather something new, which in terms of both culture and theology included and transcended both. Therefore it could be presented best only in syncretistic, parallel terms. Needless to say, the audience for such a theology of public, discernible Christian hegemony consisted of the growing Eastern urban communities. For him, in contrast to Origen, the theological locus was not the slow spiritual transformation of the individual, but the public growth of the *ecclesia*, which in its sins and social ascendancy revealed the shape and purpose of divine will for late Roman society:

> And I myself when I turn the words over in my own mind find it in no persuasive power, no dignity, no credibility, not even enough plausibility to convince one of the most foolish. But when I turn my eyes to the evidence of the power of the Word, the many it has won, the large churches which have been founded . . . not in obscure and unknown places, but in the most noble cities, royal Rome, Alexandria, Antioch, throughout the whole of Egypt, Libya, Europe and Asia, and in villages and in the country and among all peoples, I am forced to turn again to search for their causes. (*DE* 3. 7 (*GCS* 144))

EUSEBIUS AS THEOLOGIAN

Divine nature and will

Drawing on Scripture, earlier Christian authors, and contemporary philosophy, Eusebius sought to present an idea of divinity which was both universal and convincing. Although his discussions were based on traditional formulas, he was not a slavish imitator of either Origen or earlier Christian Platonist writers. In his explanation and use of common ideas such as *creatio ex nihilo*, divine participation, and the generation of the Son, he made significant modifications which, if not always successful or lucid, point to the concerns of philosophy and theology in the early fourth century. By highlighting God's sovereignty and will, with an emphasis on power as part of divine transcendence, Eusebius was able to outline a comprehensive theology of history, as well as to claim the God of Greek philosophy as the God of the Christian community.

The fundamental premiss of Eusebius's theology, as of earlier Christian apologetics, was Christian monotheism. He claimed repeatedly that a doctrine of one transcendent creator was common to both Greeks and Hebrews, quoting philosophers and prophets to prove the antiquity, universality, and hence truth of this idea of divinity.[26] Even peoples without culture shared this belief on a primitive level, because they experienced the power and will of God in nature; this was the earliest religion.[27] The wonder and truth of Christianity consisted in this transcendent deity's will and love becoming evident and accessible through the life and teaching of Jesus, the incarnate Word of God. At the beginning of the *Preparation of the Gospel*, Eusebius presents the good news of the gospel as the revelation of true religion in a blend of scriptural and philosophical terms:

> What can be more blessed than this excellent and happy friendship with God? Does he not give both life and light and truth and all good things to all? Is he not the cause of the being and the life of all things? . . . [the Word] urges all to hasten and accept the gift with eagerness: Greeks and barbarians alike; men, women, and children; rich and poor; wise and simple; even slaves. (*PE* 1. 1 (*GCS* 43. 1. 6))

Because of such passages, Eusebius's theology has been described at times as an essentially rational religion of monotheism and morality.[28] He happily equated the highest principle of the Platonic philosophers with the Creator of the Hebrew scriptures and the Father of Jesus, finding congruence between Plato, Plotinus, and Numenius on the one hand and the Gospel of John, Isaiah, and Genesis on the other. For example, the common use of ἀγαθός to describe God was treated as reason enough to assume a unity of perspective, if not a common

[26] *DE* 1. 5 (*GCS* 3. 20), 4. 1 (*GCS* 23. 150); *HE* 1. 2. 4; *PE* 11. 9 (*GCS* 43. 2. 24), 13. 3 (*GCS* 43. 2. 167).

[27] *DE* 8 praef. (*GCS* 23. 350); cf. Chesnut, *First Christian Histories*, 69 f., on Eusebius's picture of the evolution of religion.

[28] Young, *Nicaea to Chalcedon*, 18; Chesnut, *First Christian Histories*, 109; D. S. Wallace-Hadrill, *Eusebius of Caesarea* (Westminster, 1961), 196.

tie.[29] Like other apologists, he argued that Christianity made biblical ideas of creation intelligible and congruent with the best of Greek physics; the structure of a transcendent deity found an echo in the Platonic philosophers; the Christian idea of providence and free will refuted ancient fatalism and Stoicism.[30] In these arguments Eusebius's Christian scholasticism is evident; he enjoys mingling philosophical descriptions with biblical references with seemingly little or no critical theological analysis.

On the other hand, his presentation of the power and transcendence of God, as shown in miracles and martyrdom, extends beyond mere intellectual allure. He affirms a unique, universal transformation of creation through the power of divine will. As a historian and an eyewitness, he is careful to record and present the evidence of power which Christians alone possess because of their faith and knowledge of the true God. The reality of miracles, inexplicable courage in the face of persecution, and the rapid growth of the Church among disparate social classes testify to the unique work of divine power among Christians. The revelation of this one true God in Christ has brought history and reality into line with the divine will, and, as a consequence, the world abounds in evidence of the divine presence. Thus, descriptions of divine power are not merely impersonal reverberations of a higher order, but witness to the actual presence of God in history and the means of divinization; the power of God in Christ and in the disciples is a sign of true divinity.[31] Recognition of active divine power is therefore the first step to the knowledge of God, and constitutes the beginning of religion for anyone.[32] This testimony to the visible reality of divine power found throughout Eusebius's writings was of course an important pivot between Christian and Graeco-Roman piety; divine intervention was proof of authentic religion to everyone, particularly in the early fourth century. Yet, what only a few persons experienced intermittently in the Greek tradition, Christians claimed for all

[29] *PE* 11. 21 (*GCS* 43. 2. 47f.) cites Ps. 106: 1, Matt. 19: 7, *Timaeus*, 29F, *Republic*, 508B.

[30] *PE* 11. 7–14 (*GCS* 43. 2. 21–34), 11. 14–16 (*GCS* 43. 2. 35–9), 6. 6 (*GCS* 43. 1. 299f.).

[31] *Hier.* 4; *DE* 3. 6 (*GCS* 23. 133), 1. 1 (*GCS* 23. 5), 3. 7 (*GCS* 23. 141).

[32] *DE* 8 praef. (*GCS* 23. 350).

people.[33] Eusebius's monotheism therefore rested squarely on the Greek intellectual and religious tradition and the testimony of the Christian community in Scripture and experience.

To describe divine being and activity, Eusebius therefore uses a combination of terms from Scripture, Judaeo-Christian Platonism, and contemporary philosophy. The most common terms of reference were to divine being as first, perfect cause: *ἀγαθός, παμβασιλεύς, ἀρχή καί πηγή, ὁ πρώτος θεός*.[34] This description is rounded out by certain standard descriptions of transcendent simplicity: *μόνος, ἐπέκεινα τῶν ὀλῶν*.[35] Divine essence is therefore utterly transcendent and unique as the source of all being. All existence and, indeed, the Logos came from the divine being. *ὁ ὤν* defines God as the singular source of all creation.[36] In his letter to the bishop of Alexandria in support of Arius, Eusebius explained this term almost as a school lesson to chide Alexander: *ὁ ὤν* must define the Father as the unique and distinct source of all that exists.[37]

This emphasis on God as source, or *ἀρχή*, is repeated in his most common description of divine being as *ἀγένητος*.[38] Although he does not use the word in a technical way, he nevertheless underlines the singularity of God's creative essence and power by this term. Indeed, God as creator is the 'scriptural', or Hebrew, definition of God, by contrast with 'philosophical'

[33] *DE* 3. 6 (*GCS* 23. 135–6), women and children are said to outstrip Plato. Contrast this with the warnings and exhortations given to Marcella by Porphyry in *Ep. Ad Marcellam*, 7–10. R. MacMullen discusses the function of wonder and power in religion and conversion in *Christianizing the Roman Empire AD 100–400* (New Haven, Conn., 1984), 108.

[34] For use of *ἀγαθός*, see *DE* 4. 1 (*GCS* 23. 150); *PE* 1. 1 (*GCS* 43. 1. 6). For *παμβασιλεύς*, see *PE* 4. 1 (*GCS* 43. 1. 173); *DE* 5. 1 (*GCS* 23. 214). For *ἀρχή καὶ πηγη*, see *DE* 4. 3 (*GCS* 23. 153). For *ὁ πρώτος θεός*, see *DE* 5. 4 (*GCS* 23. 228); *HE* 1. 2. 8.

[35] For use of *μόνος*, see *DE* 4. 6 (*GCS* 23. 158), 4. 3 (*GCS* 23. 152). For *ἐπέκεινα τῶν ὀλῶν*, see *PE* 11. 21 (*GCS* 43. 2. 48), 7. 15 (*GCS* 43. 1. 393); *DE* 4. 7 (*GCS* 23. 161); *CM* 1. 4 (*GCS* 14. 23).

[36] See *θέλω* and *ὁ ὤν* (Exod. 3: 14) in *DE* 4. 1 (*GCS* 23. 151); in the same passage he also uses Rom. 11: 36; cf. *ET* 2. 20 (*GCS* 14. 129); cf. *PE* 11. 9 (*GCS* 43. 2. 24–5).

[37] *Ur.* 7. 4 (ed. Opitz, 15): *πάλιν αὐτοὺς ᾐτιῶ λέγοντας ὅτι >> ὁ ὤν τὸν μὴ ὄντα γέννησε <<. θαμάζω δέ, εἰ δύναται τις ἄλλως εἰπεῖν;*

[38] e.g. *DE* 4. 3 (*GCS* 23. 153), 4. 1 (*GCS* 23. 150); *PE* 4. 5 (*GCS* 43. 1. 175); *HE* 1. 2. 8. Cf. G. L. Prestige, '*Ἀγέν[ν]ητος* and *γεν[ν]ητός* and Kindred Words in Eusebius and the Early Arians', *JTS* 24 (1923), 486–96.

descriptions.[39] In other words, God is the creator of all that exists, and his essence is unique as ἄναρχος and ἀγενητος.[40] For Eusebius ἀγένητος was both traditional and essential to underline the creative power and uniqueness of divine nature. Throughout his works he defines divine nature largely with reference to causality: the Father alone has his existence from himself, and his nature is therefore unique and transcendent.[41]

Like other pre-Nicene theologians, Eusebius accepts that God as ἀγένητος cannot by nature appear to humanity.[42] But, this does not mean that God is distant; rather, God is continually present and active through his power and will in shaping human history. As in earlier Christian authors, his descriptions of divine transcendence are often affirmations of God's power.[43] The very transcendence of God is what allows divine intervention at any level of reality. Eusebius accepts the philosophical descriptions of transcendence as necessary and worthy of God; but as a theologian he uses them to safeguard the absolute power of divine will throughout existence.

His views concerning transcendence and power may be clarified by a closer examination of the relation between will and nature in his theology. In the *Demonstration of the Gospel* Eusebius discusses divine will and creation thus:

> For because God wills it is the sole cause of all things that exist coming into being and continuing to be. For it comes by his will and he wills it, because he happens to be good by nature [θέλοντος γάρ ἔστιν, θέλει δέ, ὅτι τὴν φύσιν ἀγαθὸς ὢν τυγχάνει]. For nothing else is essential by nature to good except to will what is good. And what he wills, he can do. Therefore, having both will and power he has established . . . everything beautiful and useful in the visible and invisible world, making his own will and power a kind of material substratum of the beginning and constitution of the universe, so that it is no longer reasonable to say that anything that exists must have come from the non-existent [τὸ μὴ ὄν] for that which came from the non-existent would not be anything. . . .

[39] *PE* 13. 15 (*GCS* 43. 2. 232); cf. *PE* 11. 21 (*GCS* 43. 2. 48), 7. 18 (*GCS* 43. 1. 400).

[40] *DE* 4. 3 (*GCS* 23. 153); *CM* 1. 1 (*GCS* 14. 4); *ET* 2. 23 (*GCS* 14. 133).

[41] *DE* 4. 6 (*GCS* 23. 158); cf. *DE* 4. 1 (*GCS* 23. 151).

[42] *HE* 1. 2. 8; cf. Justin, *Dial.* 56, 127; Irenaeus, *Haer.* 3. 9. 1. On transcendence, see A. Weber, Ἀρχή. *Ein Beitrag zur Christologie des Eusebius von Caesarea* (Rome, 1964), 32–40; Ricken, 'Logoslehre', 344–7.

[43] *DE* 1. 5 (*GCS* 23. 22), 4. 6 (*GCS* 23. 158), 4. 7 (*GCS* 23. 161). Cf. Ricken, 'Rezeption', 331.

Everything that has been created or now exists derives its being from the one, the only existent and pre-existent being, who also said, 'I am the existent [εγώ εἰμι ὁ ὤν]. (*DE* 1. 4 (*GCS* 23. 151))

A parallel to this passage was written twenty years later in his *In Praise of Constantine*: 'θέλοντος γὰρ ἐστὶν ὅ τι καὶ ἐστὶν καὶ μὴ θέλοντος οὐκ ἐστὶν, θέλει δὲ ἀγαθὰ παντὰ, ὅτι δὴ καὶ αὐτοάγαθον τήν οὐσίαν τυγχάνει.'[44]

A careful look at these passages reveals much about Eusebius's Christian Platonism. Although this description of creation is clearly informed by the *Timaeus*, Eusebius has modified philosophical commonplaces about divine goodness and will in several important ways. First, he has altered the Platonic definition of God as essentially good to enhance the sense of divine freedom. The presence of τυγχάνω with θέλω in both passages suggests his awareness of the problem of necessity not only in regard to God's creative activity, but also in regard to nature. This does not mean that God could choose to be other than good—indeed, the goodness of God is often cited by Eusebius as a guarantee of the goodness of creation and as a firm parallel to Platonism—but the theological concept of God's gracious goodness is conveyed in the curious voluntaristic description 'ἀγαθῷ δὲ οὐδέν, ὅτι μὴ τὸ θέλειν τὰ ἀγαθὰ προσφυές.' This statement has a general resemblance to Plotinus's discussion of the freedom of the One; but, as discussed above, Plotinus's concern is with the inner life, which is ultimately beyond freedom or necessity. Eusebius's focus on the outward power and will of divine goodness resembles that found in other late third- and early fourth-century rebuttals of co-existence and fatalism: God's eternal nature does not curtail choice or power in creation.[45]

Eusebius went on to assert the importance of divine will as the critical link between God's essence and power, if they can be separated: 'θέλων δὲ ταῦτα καὶ δύναται'. This is reminiscent of

[44] *LC* 12 (*GCS* 7. 1. 230).

[45] Earlier Christian authors, such as Clement in *Strom.* 2. 16, 6. 16 and Hippolytus in *Con Noet.* 8. 15, had modified the idea of essential goodness of divine nature towards a more voluntarist understanding. For the third century, see Methodius, *Creat.* 2–6, and G. C. Stead, 'The Platonism of Arius', *JTS* NS 15 (1964), 30. On the use of will against the Manichees, see Alexander of Lycopolis, *Critique of the Doctrines of Manichaeus* 17 (ed. Mansfeld, 80–1). For Porphyry's defence of divine power *contra* Plotinus, see above, p. 14–15. On Eusebius, see also Stead, *Divine Substance*, 231–2.

Origen's argument in *On First Principles* 1. 2. 6, but the voluntaristic element is greater. In both passages the emphasis on the goodness of God is matched by an affirmation of God's will and power. In the later discussion Eusebius is happy to anchor this in a term of utter transcendence, *αὐτοάγαθον*. In the system of Numenius this term refers to the immovable character and absolute transcendence of the first principle, which is explicitly beyond will.[46] Eusebius thus combines the strongest contemporary metaphysical definitions of transcendence and voluntarism to assert God's absolute creative power.

The difficulties of Eusebius's synthesis are evident in the curious description of *creatio ex nihilo* in this passage. Because God is the source of all that exists, '*μηκέτι εὐλόγως φάναι δεῖν, ἐξ οὐκ ὄντων εἶναί τι τῶν ὄντων.*' With the help of Exodus 3: 14 and a conflation of Romans 11: 36 and Acts 17: 28, Eusebius describes the radical contingency of all creation, yet tries to maintain its intimacy with its creator by the sheer fact of existence. Although the precise nature of the relation is difficult to make out in this context, it can hardly be a literal statement, given his defence of the uniqueness of divine being. Indeed, it seems similar to Porphyry's description of all existence outside the first principle as 'non-existent' relative to the absolute being of the first cause.[47] Thus Eusebius, like Irenaeus, describes God's will as the substance of the universe (*καὶ συστάσεως τὴν ἑαυτοῦ βουλὴν καὶ δύναμιν προβεβλημένος*). Both statements signify the radical dependence of all existence on God's will and power, rather than a literal sharing in divine essence.[48]

In these passages concerning creation, Eusebius asserts both God's absolute ontological independence by nature and his intimate involvement with creation by will. The theological thrust here is divine will as personal purpose rather than abstract or automatic power. Although he retains terminology from the dynamic emanationism of Origen, he defines the difference between divine and created nature more sharply as *ἀγένητος* and *γένητος*. In creation, as in soteriology, he does not imply a chain of being that is shared in some way by God and humanity. The

[46] Fr. 3, 16 (ed. des Places).

[47] See Porphyry, Fr. 2, in Hadot, 'La Métaphysique de Porphyre', 133–5; cf. Stead, 'Platonism of Arius', 25–6.

[48] See also discussion in Weber, Ἀρχή, 37.

essence of the first principle is absolutely unique, and only by will is it linked to creation. However, the sheer dependence of creation is matched by the sheer power of divine will.

Eusebius emphasized a similar ambiguity in the transcendence of divine nature and the power of will in the generation of the Son. As will be discussed more fully below in regard to Christology, he modifies the traditional images of generation from Wisdom so as to accentuate the choice of the Father, rather than implying any sort of automatic causality:

> The ray does not shine forth from the light by its choice, but because it is an inseparable accident of its essence; but the Son is the image of the Father by intention [γνώμην] and choice [προαίρεσις]. (*DE* 4.3 (*GCS* 23. 153))

All existence, therefore, including the Son is thrown into relation with God by will, as iron bits to a magnet. The Son, as divine, has a unique relationship with the Father, but a separate existence.[49] These modifications in the direction of voluntarism in creation and generation of the Son seem to stem from a concern found also in Methodius's critique of eternal creation in Origen and contemporary polemics against the Manichees.

Not surprisingly, these ideas led to his sympathies being inclined initially towards Arius during the Nicene controversy. His two letters of support offered scriptural and logical proof of the primacy and uniqueness of the Father's essence. First, on the basis of John 14: 28 and the logic of causality, the Father was the only true God; second, the Son, precisely because he is Son, had to be separate, if unchanging (*ἄτρεπτον καὶ τέλειον κτίσμα τοῦ θεοῦ*).[50] Although Eusebius accepted the Nicene formula, he did not relinquish his previous conviction regarding the uniqueness of the Father's essence as *ἀγένητος*; and his explanation of *ὁμοούσιος* followed his own pre-Nicene account of the Father as the direct cause of the Son, rather than allowing any sense of identity of essence between them. Even so, the alleged Arian definition of the origin of the Son from nothing by the will of the Father was not acceptable to him either.[51] He notes in his

[49] See θέλω and Exod. 3: 14 in *DE* 4. 1 (*GCS* 23. 151); *ET* 1. 9 (*GCS* 14. 67). See also Stead, *Divine Substance*, 240.

[50] *Ur*. 3. 6 (ed. Opitz, 4–5), 7 (ed. Opitz, 14–15).

[51] *Ur*. 22. 12–14 (ed. Opitz, 45–6). For a sympathetic account of Eusebius's letter, see Stead, *Divine Substance*, 240.

account of Nicaea that Constantine agreed with his interpretation of eternal generation: that the Father always possessed the potential to beget the Son. This account suggests Eusebius's earlier description of the transcendence and power of God as grounded in his unique essence and his relation to the world by his will alone. In his later works Eusebius continued to defend the biblical and necessary primacy of the Father, which, as he claimed, was thoroughly traditional.

Eusebius's rhetorical skill in explaining the transcendence and power of God is exhibited in his *Tricennial Orations*. Rather than being evidence of Eusebius's uncritical use of Middle Platonism, as claimed by Ricken, this work employs a deliberate style of address in which little Christian language is used, so as not to offend non-Christian sensibilities.[52] The language of the orations, therefore, should not be taken as typical of Eusebius's theology, but rather as evidence of his knowledge and use of philosophy for the benefit of the imperial court. In his *In Praise of Constantine* Eusebius's knowledge of Numenius is well exhibited, and is perhaps used to appeal to the emperor's own Middle Platonic and even Numenian orientation.[53] The Father is therefore accorded the transcendent descriptions of Platonism, blended with opaque references to Scripture. As in his other works, Eusebius uses the traditional transcendence of the Greek deity to highlight the gracious power of the Christian God, and continues to appeal to evidence of divine power as proof of Christianity. Thus, even in his most philosophical work, Eusebius emphasizes the power and accessibility of the Father, along with the utter transcendence of the divine essence.

Like earlier apologists, he affirms the power of divine providence over against contemporary fatalism as an essential Christian teaching: creation by an all-powerful God is the guarantee of human freedom because the very order of existence is in accordance with divine will:

> But through all existing things, both those that occur through us, and by their own causes, and accidentally from without, and those of nature, one all powerful and almighty providence [πρόνοια] goes throughout the

[52] Drake, *In Praise of Constantine*, 28–9; Ricken, 'Rezeption', 326, 331.

[53] Barnes, *Constantine and Eusebius*, 74–5; idem, 'The Emperor Constantine's Good Friday Sermon', *JTS*, NS 27 (1976), 414–23; Rist, 'Basil's "Neoplatonism"', 155–8.

whole . . . guiding it through obedience and altering many natural things to suit the occasion, working with and co-operating with our own wills. (*PE* 6. 6 (*GCS* 43. 1. 307))[54]

Eusebius was determined to prove this by the congruence between the divine plan and the progress of history; he therefore translated Origen's optimistic cosmology into universal history. The will of God thus works with individual freedom through accidental events (συμβεβηκός) to bring salvation history to completion.[55] *Πρόνοια* is therefore *δύναμις* which works throughout material being and history to transform society in accordance with God's desire.[56] The persecution of the Church, the endurance of the martyrs, and the conversion of the emperors all reveal the pedagogical plan of God: 'This must have been the work of almighty God, the submission of the enemies of the Word to a greater fear of the supreme ruler. For God wills it daily to advance and spread among his people' (*DE* 3. 7 (*GCS* 23. 146)).

Eusebius's assertion of the transcendence of God in accordance with contemporary philosophy is thus matched by an equally strong defence of God's will and power. His emphasis on creation and his definition of God as *ἀγέννητος* are indeed traditionally Christian and based on certain scriptural ideas. Given this sharp defence of the freedom of the divine will, it is difficult to continue to assert an uncritical dependence on philosophy or, as Ricken has maintained, the assignment of equal weight to philosophy and Christianity in Eusebius's apologetic theology.[57] Clearly, Eusebius accepted and used the structure of later Platonism to understand the transcendence and power of divine being and its relation to the world. However, his modification of being and will pushed divine activity and cosmology in a different direction from the transcendent ontology of Numenius or Plotinus. The Christian God, if ontologically transcendent,

[54] Cf. *DE* 3. 3 (*GCS* 23. 112f.); *PE* 6. 6 (*GCS* 43. 1. 304), 6. 6 (*GCS* 43. 1. 307f., 310f.).

[55] *PE* 6. 6 (*GCS* 43. 1. 310f.); *HE* 10. 4. 20–1; *Hier.* 42. Cf. Chesnut, *First Church Histories*, 41–50.

[56] Origen defines *πρόνοια* as *δύναμις* in *Cel.* 6. 71 (*SC* 147. 358). On *πρόνοια* and the activities of Christ in Eusebius, see *ET* 1. 12 (*GCS* 14. 72), 3. 2 (*GCS* 14. 142).

[57] Ricken, 'Rezeption', 333; see also his discussion of Eusebius, Plotinus, and Numenius based on *PE* 11. 9, ibid. 324–7.

remains intimately involved in the creation through his power and will, and particularly through the activity of the Logos. Increased transcendence is therefore matched by increased voluntarism and power in Eusebius's definition of divinity.

Human nature and will

Like Origen, Eusebius believed that God's purpose for creation would eventually be fulfilled with the co-operation of human freedom; history was the material stage for divine pedagogy and human progress towards union with God. With Justin and Irenaeus, Eusebius shared a rational optimism about the moral and spiritual progress of humanity under the instruction of the divine will. However, his anthropology has more often been linked to 'exterior' philosophical rationalism and morality than to the 'internal' transformations of grace.[58] Yet, as with his doctrine of God, his apologetic use of philosophical categories may mask his theological intentions, which are revealed not by his vocabulary, but in the relationship between God and humanity through the Logos. Not merely in universal moralism or impassible virtue, but in the martyrdoms and the triumph of the Church, lay the evidence of divine power to change and transform the lives of all people.

The gracious intention of God in creating humanity in the divine image as good, rational, and free is repeatedly praised by Eusebius:

> For God, the only good, and the source and cause of everything good, willed to make many partakers of his riches, creating the whole rational creation, incorporeal, intelligent and divine powers, angels and archangels, pure spirits, and human souls endowed with unfettered and free self-determination about the good [*ἄφετον καὶ ἐλεύθερον τῆς αὐθεκουσίου*], and give to them whatever bodily organs they were to possess, suitable to the variety of their lives, with countries and places natural to all. For those who had remained good, he gave the best places, and to those who did not he gave fit abodes, places of discipline for their willfulness [*ῥοπῇ*]. (*DE* 4. 1 (*GCS* 23. 150))

Following Origen in his use of degrees of moral goodness to understand spiritual and social diversity, Eusebius, overtly at

[58] Young, *Nicaea to Chalcedon*, 18; Chesnut, *First Christian Histories*, 109.

least, attributes this to divine foresight rather than a pre-existent Fall. Given the criticism directed against Origen's use of this idea, it is not surprising to find it modified in Eusebius. Thus he describes the deliberate creation of humanity in a material state, even though the soul was created first as immortal. Since he gives no reason for material creation other than the difference between humans and angels, one may suspect that he was ignoring the Origenist question entirely and describing the accepted structure of existence, rather than enquiring into its cause.[59] In view of his historical interests, it is possible that Eusebius may have been following Methodius and Irenaeus in affirming the material creation on its own. However, like Origen, Eusebius considered material life as a planned stage, an 'infantile constitution' which was eventually surpassed in spiritual perfection:

> [If] this brother of the divine hosts, and of angels which are in heaven, had been duly led by his nature, and had from ancient time adhered to the divine law, he would indeed have been freed from this earthly and corruptible life and continued in his convention on earth . . . as in a state of migration (with Phil. 3: 20, Heb. 12: 22). (*Theo.* 1. 69 (*GCS* 11. 70))

Humanity, therefore, according to Eusebius, was created in the image of God with an immortal soul and a material body. Compared with the splendour of God, human nature, though good, was contingent, multiple, and weak.[60] The tension between body and soul led Eusebius to speak of a 'double nature', making one at once 'slave and free':

> Humanity is not of one only begotten thing nor from one sort of nature, but is from two opposites, body and soul. . . . One is irrational and the other rational, one mortal and the other immortal. . . . So this double nature orders its life in a twofold and diverse manner, at one time enslaved to the body and at another welcoming with the diviner part its freedom. Thus, one is both slave and free, being so from God for God's own purposes. (*PE* 6. 6 (*GCS* 43. 2. 303))

In the soul one finds the image of God: that is, natural law, reason, and free will, which enable a person to live a godly life.[61] The goal of human life is to follow this innate good nature and

[59] *Theo.* 1. 44 (*GCS* 11. 2. 98–9); *DE* 3. 3 (*GCS* 23. 113); *PE* 7. 18 (*GCS* 43. 1. 400). Chesnut defends pre-existence in Eusebius (*First Christian Histories*, 68–9).

[60] *DE* 4. 6 (*GCS* 23. 158).

[61] *DE* 3. 3 (*GCS* 23. 112).

rise beyond created limits to become a partaker of the divine spirit and converse with the angels.[62] As in earlier Christian anthropology, free will is therefore central to human nature and salvation. All creatures share one common, good nature as created by God. This is a source of hope, since, in opposition to Gnostic or Manichaean anthropology, everyone is created in the image of God. For Eusebius, as for Origen, the created goodness of humanity is an internal lodestone towards God. The fact of being created in God's image and in God's care means that humanity, even in its freedom, possesses an inclination towards goodness:

For God made neither nature nor the substance of the soul evil, for a good being may not create anything but what is good. Everything is according to nature good, and every rational soul has by nature the good of free will [τὸ αὐτεξουσίον ἀγαθόν] which is for choosing what is good. (*PE* 6. 6 (*GCS* 43. 1. 308))

Free will is thus a positive gift, rather than a neutral trait, which safeguards the justice of God and moral responsibility. The nature of the soul as created by God is to incline towards the rational and the good, just as the nature of the body is to incline towards the sensual. In accordance with its constitution, the soul retains the power to be free to be virtuous:

For although a myriad of outside and accidental impediments oppose both the body's nature and the independent efforts of our will, nevertheless the free virtue in the soul persists, revealing that the choice of the good, which lies within us, is unconquerable and invincible. (*PE* 6. 6 (*GCS* 43. 1. 310))

In accordance with earlier Christian tradition, Eusebius defines the freedom and constitution of the soul as self-moved in a positive sense, rather than undetermined or evil. In his first work Eusebius describes the soul as immortal, because it is always moving (ἀεικίνητος) in accordance with choice, not necessity.[63] The soul is self-moved (αὐτεξούσιος) and indeed, using a recently coined term also found in Methodius, self-willed (αὐτοπροαίρεσις).[64] The body is subject to natural laws, but the will

[62] *Theo.* 1. 38 (*GCS* 11. 2. 56–7); *HE* 10. 4. 55. [63] *Hier.* 41.
[64] *PE* 6. 6 (*GCS* 43. 1. 301); *DE* 4. 6 (*GCS* 23. 160), 9. 4 (*GCS* 23. 412). See Methodius, *Res.* 1. 38.

should control and train the body; however, the body alone is not the source of sin, because the soul can also be foolish and led astray.[65] Human error or carelessness, rather than the stars or God, is the source of evil and distance from God. Using standard philosophical arguments against fate, Eusebius goes on to outline not only intellectual carelessness or bodily distraction as allowing evil, but also external temptations, the difficulty of civil and social life, and the effect of others on human will.[66] Yet, for the Christian, hope lies in the pervasiveness of God's will, which is continually co-operating with human will, even in these distracting material and social circumstances. God is the creator of human power, human nature, and all accidents.[67] Self-motion also ensures the moral responsibility of humans and the justice of God: humans must be free to save themselves; the options may be presented by God, but it is the individual who chooses.[68] The universe, therefore, is constructed of a multiplicity of self-moving, continually choosing individuals guided by a beneficent providence.

Estrangement from God is the result of wrong choice, the possibility of which is part of created freedom, yet contrary to one's created good nature. Thus Eusebius stresses the abuse of freedom in the turning from God toward self:

> It [evil] will be found solely in the self-determined motion [*αὐτο-προαιρέτῳ*] of the soul, and in this . . . when it departs from the king's highway and turns by its own decision against nature, being its own master. For the soul has obtained freedom and self-mastery from God, and assumed the determination of its own motion; but the divine law united with it by nature, like a beacon and a star, calls to it with a voice from within and says, 'You shall walk the king's highway, you shall not turn aside to the right hand nor to the left.' (*PE* 6. 6 (*GCS* 43. 1. 307))

Hence sin consists in wrong motion away from God. The contrast between human unfaithfulness and the constancy of stars or animals is part of the reality of historical progress, rather than an association of freedom with evil as Stead has suggested.[69] Eusebius does not pessimistically equate motion itself with

[65] *PE* 6. 6 (*GCS* 43. 1. 303). [66] *PE* 6. 6 (*GCS* 43. 1. 305).
[67] *PE* 6. 6 (*GCS* 43. 1. 309); *DE* 4. 1 (*GCS* 23. 150); cf. Chesnut, *First Christian Histories*, 45 f.
[68] *Theo.* 1. 4 (*GCS* 11. 2. 59); *PE* 10. 4 (*GCS* 43. 1. 572).
[69] *DE* 4. 9 (*GCS* 23. 163); Stead, 'Freedom of Will', 250.

created instability, but sees it as a free gift which has been used wrongly. Virtue, as well as vice, requires motion or effort, since the innate nature of the soul is not static.

Human freedom and divine providence are central not only to Eusebius's argument against fatalism, but also to his account of Christianity's triumph in history. Eusebius traces the historical lapse of Adam from a primitive worship of God in 'simplicity of mind' to dissolute idol worship.[70] This idol worship Eusebius sees as the historical manifestation of misused freedom; in a way that resembles Irenaeus's theology of history, Eusebius views polytheism as pedagogical history derailed. Humanity has ignored divine instruction for 'irrational itchings and delusions of childhood'.[71] Having fallen away from worship of the true God, humanity has become ignorant of its own nature and destiny:

> becoming subject to disease and madness, so that they knew neither God as their Father nor the proper essence of their own spiritual nature, nor God's providence which preserves the whole, but had almost become irrational animals. (*PE* 13. 3 (*GCS* 43. 2. 176))

As a result, human freedom has become vulnerable to the lure of demons; human nature is 'dead' from the Fall.[72]

The incarnation of the Logos has restored communion—that is, knowledge and power—with God, and has thus restored true power and freedom to humanity. The focus for Eusebius in this process is communal and historical. In the *Ecclesiastical History* he lays out an elaborate parallel between the redemption of the soul and the redemption of the church. Both the soul and the church were dead, corpses; in each case this occurred because of her free choice (*ἐξ αὐτεξουσίου αἱρέσεως*), which led to her being raped by demons, profaned, and even buried. The Word subsequently cleansed and gave life to the soul/church again through redemptive power and teaching:

> And by this way those souls which shortly before had been fouled and covered by impious commands he cleaned by pickaxes and mattocks, that is by the striking words of his teaching. He made bright and clear the minds of all. (*HE* 10. 4. 61)

[70] *Theo.* 1. 64 (*GCS* 11. 2. 69). On irrationality, see Chesnut, *First Christian Histories*, 97 f.

[71] *Theo.* 3. 62 (*GCS* 11. 2. 160); *DE* 4. 7 (*GCS* 23. 160), 4. 8 (*GCS* 23. 162).

[72] *HE* 8. 7. 2, 8. 9. 8, 10. 4. 61.

The locus of salvation for Eusebius is the restoration of the soul, which may then train the body. However, the transforming power of Christian teaching through the Word and its disciples does not consist only in intellectual discernment or moral purity, but in the communication of divine life to the lost individual. His examples of this transformation are usually communal. The power of God is best exhibited in the Christian community.

> They found one and the same destiny, being brought under one word and teaching, and revealing one mind [γνώμην], and one will [προαίρεσις], and one virtue of soul: accepting the same life, loving the same Word, and enduring with love the same sufferings for their steady piety. . . . Young and old together, of every age, male and female, barbarians, slaves, free, learned and uneducated . . . prefer a philosophy that consists not in words, but in acts. (*PE* 6. 6 (*GCS* 43. 1. 311–12))

The evidence of divine power and transformation lies precisely in the fact that ordinary people are able to live extraordinary lives; Christian teaching as the communication of redemptive power 'consists not in words but in acts'. Thus the martyrs showed the strength of perfect freedom and divine will by their endurance and voluntary sacrifice.[73] Within the community Eusebius describes divine power operating at two levels in Christian life: in the lives of ascetics who devote all their energies to the worship of God and 'in mind and spirit pass to heaven . . . the soul purified in disposition' and in the lives of more humble people who are engaged in everyday life, but with a secondary piety nourished by religious discipline and liturgy.[74] Even though Eusebius, like most of his contemporaries, views the ascetic life as the higher form of Christian dedication, akin to philosophical perfection, he more commonly boasts of the universality of virtue and piety brought by Christianity: rich, poor, slave, free, man, woman, barbarian, and Greek in their worship of one God, their holy living, and voluntary courage in the face of death testify to the truth of Christian teaching. Christian virtue consists in a steadfast communion with God.

Eusebius's vision of Christian life is thus one which fulfils the highest philosophical ideals in all persons and re-establishes true

[73] *DE* 1. 8 (*GCS* 23. 39).

[74] *DE* 4. 10 (*GCS* 23. 167), 4. 13 (*GCS* 28. 172). On Eusebius's vision of the cosmic liturgy, see Chesnut, *First Christian Histories*, 107–9.

communion with God. He therefore celebrates morality and piety, because these traits witness to the communal restoration of society. On one level Eusebius seems to be equating Christian life with classical moralism—hence the charges of an 'external' soteriology. This is in part due to the apologetic nature of his work, which, as a standard theme, celebrates the ethical superiority of simple Christians. Yet, woven into this is the transformation of humanity by divine power towards incorruptibility and union with God.[75] If the main instrument for Eusebius is teaching, an instrument which brings the power of God to the soul and the body, it is not a lower-grade philosophical instruction, but a teaching that looks to purification of the soul in order to master the body and live in accordance with divine will. Like Origen, Eusebius affirms a teaching which also communicates the qualities of wisdom and deification to the believer as an adopted child of God. Thus moral behaviour is not external, but a sign of inner transformation and the metaphysical link to reception of continued divine power and to communion with God.

Humanity has been created for friendship with God in material life; therefore right behaviour promotes a living, growing relationship between God and humanity. Bodily weakness may hinder human life; but it will not thwart God's ultimate purpose in history, which will be accomplished by the education and transformation of the will. Material existence and hence history constitute the proper, intended place for human development, and are in fact, the place where progress can be observed. Although Eusebius defends individual responsibility over against fatalism, he focuses on the communal, social aspects of soteriology, rather than individual experience. Salvation history as revealed in the lives of the martyrs and ascetics proves that free obedience to the Logos results in unchanging, cheerful adhesion to God and eventual divinization, not merely of the individual, but of the entire society as well.

Christology

By later standards Eusebius's subordinationist and cosmological Christology seems to vitiate any coherent Christianity. This Logos–sarx Christology did not do justice to either nature.

[75] *HE* 10. 4. 46; *DE* 9. 13 (*GCS* 23. 432); *ET* 3. 18 (*GCS* 14. 179).

Because the Son was not fully divine, he was unable to divinize, and because the flesh was only an instrument, genuine human obedience was impossible. Eusebius's understanding of the Incarnation was therefore determined by the problems of cosmological mediation.[76] However, a closer look at his patterns of exegesis and the function of his Christological model may help us reconstruct his theological intentions concerning mediation more sympathetically. Christ as divine Logos and human teacher fulfilled the transcendent Father's will in both creation and history, bringing order and power on both levels. The cosmological and historical activities of the Son were two sides of the same coin: 'Our Saviour and Lord, the Word of God himself, was known between two lives . . . one life according to God, the other according to man; the one mortal, the other eternal.'[77]

The Middle Platonic origins of the cosmological Son have been well researched by Ricken.[78] Not surprisingly, close parallels exist in language and function between the second principle of Numenius, Atticus, and Albinus and the Logos of Eusebius. The latter was *ὑπουργός, δεύτερος θεός*.[79] Described as the helmsman of the universe and the demiurge, he was the creator of the noetic, sensible world, as well as the principle of order and life, like the Platonic world soul.[80] The position of the Logos as intermediary between uncreated and created substances seems to echo the position of the second god of Numenius, who protects the transcendence of the highest god by his relation to matter. The Logos as cosmological mediator does not share the first god's attributes by essence, but by participation; a separation and unity of essence exists between them; he is the power of the transcendent Father in the created realm.[81] Admittedly, these descriptions were also Origenist and Philonic, and Ricken notes that Eusebius

[76] Lienhard, '"Arian" controversy', 423; Grillmeier, *Christ in the Christian Tradition*, 177; Wallace-Hadrill, *Eusebius of Caesarea*, 120: 'His soteriology founders on the quicksand of his Christology.' On transcendence and immanence issues as fundamental, see Opitz, 'Euseb von Caesarea', 18; Ricken, 'Nikaia', 340.

[77] *DE* 6. 15 (*GCS* 23. 270).

[78] Particularly in 'Logoslehre', and 'Rezeption', 324–31.

[79] *DE* 4. 10 (*GCS* 23. 167). For *δεύτερος θεός*, see *PE* 7. 12 (*GCS* 43. 1. 388); *DE* 5 praef. (*GCS* 23. 202), 5. 1 (*GCS* 23. 214). See also Ricken, 'Logoslehre', 349–50.

[80] *DE* 4. 5 (*GCS* 23. 155), 4. 10 (*GCS* 23. 167–8); Ricken, 'Logoslehre', 348.

[81] Ricken, 'Logoslehre', 355–6; 'Rezeption', 331.

may have drawn this language from these authors.[82] However, Eusebius's use of Middle Platonic language indicates his profound acceptance of a definite ontological hierarchy, and explains his radical subordination of the Son, especially if he followed a Numenian model. Ricken concludes that such parallels do not inform us as to whether Eusebius was concerned more with historical dispensation or with problems of transcendence and immanence; but his Christology is hindered, if not determined, by the assumptions of Platonic cosmology.[83]

Although very helpful in outlining the philosophical sources of Eusebius's Christology, Ricken's interpretation has several difficulties. Many of the Middle Platonic parallels are drawn from works which were deliberately couched in philosophical language. Although Eusebius used Platonic language throughout his writings to describe the role of the Son, as discussed above, *In Praise of Constantine*, a work frequently cited by Ricken, was particularly Platonic in languange on account of its audience. Equally important, in citing the conceptual or verbal parallels, Ricken does not examine their scriptural links or theological function. Thus, he concludes that Eusebius as an apologist blended scriptural and philosophical language; yet the subordination and cosmic role of the Son were the result of philosophical influences.[84]

In his apologetic works Eusebius set out to prove from philosophy and Scripture that Jesus, the incarnate Logos, was the unique agent of the Father's will foretold by the Hebrew prophets and mirrored in Platonic writings. Hence he deliberately considered the theology of Christ from both historical and philosophical viewpoints. From either perspective Jesus was unique. This basic method was most extensively employed in the *Preparation* and *Demonstration of the Gospel*. In the latter, Eusebius outlined the two ways of examining the theology of Christ:

> Two ways of considering our Saviour Jesus Christ have been illustrated . . . the first takes us above nature and beyond it: here we define him as

[82] Ricken, 'Logoslehre', 355.

[83] Ibid. 358. In his later article he contrasted Eusebius to Athanasius as interested in cosmological mediation rather than soteriology (Ricken, 'Nikaia', 340; cf. Berkhof, *Theologie*, 62).

[84] Ricken, 'Rezeption', 333.

the only begotten Son of God or the Word who is of the essence of God, the secondary cause of everything or a spiritual substance, and the firstborn nature of God, his holy and perfect power, or the spiritual image of the unbegotten nature. The second was akin to us and more familiar: here we recognized Christ as the Word of God, proclaiming in human nature the holiness of the Father. (*DE* 5 praef. (*GCS* 23. 202))

In this passage Eusebius blends scriptural phrases with certain categories from Platonic ontology. He notes that the sources for a theology of the Son include John 1: 1, Colossians 1: 15, and 2 Corinthians 1: 24, along with the Hebrew evidence of Wisdom, Proverbs, and prophetic descriptions.[85] Thus, in this description of the Word (John 1: 1) as being from the *essence* (*οὐσιώδη τοῦ θεοῦ λόγου*) of God, the first-born (*πρωτότοκον . . . φύσιν*) *nature*, and *spiritual* (*νοερὰν εἰκόνα*) image (Col. 1: 15), he mixes biblical phrases and metaphysical language to underline the transcendence and divinity of the Son. This deliberate synthesis represents an attempt to make the divinity of the Son intelligible to educated Greeks and Christians and also to make a claim of universality for Christ's work as the divine Logos. Later, against Marcellus, Eusebius explicitly notes that the Son is not an impersonal Logos or a mere attribute of God, but that his relation to the Father, as shown in Scripture, reveals him as a living Son.[86] His Christology therefore rests on a profound and original, if not always coherent, synthesis of biblical and philosophical models of mediation.

In contrast to both contemporary philosophy and Origen, Eusebius emphasizes not only the transcendence of God, but also the will of the Father in the generation of the Son. On the one hand, he uses scriptural metaphors from Wisdom, and does not get far beyond them. Thus, in *Demonstration* 4. 3 he uses Wisdom 7: 22–5 and Hebrews 1: 3; in *Demonstration* 5. 1, Proverbs 8: 12–21 and Colossians 1: 15, leaning on the mystery of generation (Isa. 53: 8), as in the first discussion. In both cases the discussion consists mainly of an analysis of the language used in Scripture and hence the properties of light, fragrance, power, and image. On the other hand, he reflects at length on the free, volitional power of the Father in the generation of the Son:

[85] *DE* 5 praef. (*GCS* 23. 209). [86] *ET* 1. 20 (*GCS* 14. 83).

For 'he is the radiance of the eternal light and the unblurred mirror of the activity of God and the image of his goodness'. . . . But radiance is inseparable from physical light, whereas the Son exists in himself, in his own essence apart from the Father [*ὁ δ' υἱὸς ἰδίως παρὰ τὸν πατέρα καθ' ἑαυτὸν ὑφέστηκεν*]. And the ray has its own activity solely from the light, whereas the Son is something different from a channel of energy having his own being. And again the ray coexists with the light. . . But the Father precedes the Son, and has preceded him in existence for he alone is unbegotten. The one, perfect in himself, and first as Father, the cause of the Son's existence receives nothing towards the completeness of his divinity from the Son. . . . The other, as a son begotten of him, came second as a son, receiving from the Father both his being and the character of his being. And moreover the ray does not shine forth from the light by its choice, but it is an inseparable accident of its essence: but the Son is the image of the Father by intention and choice [*ὁ δὲ υἱὸς κατὰ γνώμην καὶ προαίρεσιν εἰκὼν ὑπέστη τοῦ πατρός*]. For God willed (*βουληθεὶς*) to beget a Son of the Father and establish a second light. (*DE* 4. 3 (*GCS* 23. 152–3))

In this passage each of the traditional metaphors of divine generation is qualified in such a way as to ensure the separate identity of the Son and the choice of the Father. Thus, the Son is not a channel of energy, but a being in himself; he is not a co-existent ray, but second to the unbegotten essence of the Father. As noted above, this concern with eternal co-existents and reliance on will echoes contemporary criticism of Manichaean dualism and Methodius's criticism of Origen's doctrine of eternal creation. Eusebius does not hesitate to use the language of decision (*προαίρεσις*) in generation. He is also specifically concerned to exclude any notion of emanation, even modifying the term 'fragrance', perhaps in response to Plotinus, to express an eternal, but free generation: the Son was cast forth before all times by the Father's will and power.[87]

Although Eusebius was eager to preserve the freedom of the Father in generation by the use of will as decision rather than power, he also defends the divine likeness and nature of the Son. In contrast to Arius, he maintains that the Son does not come from nothing or have a temporal beginning as a result of his having been caused.[88] Eusebius even denies any 'declension or inferi-

[87] *DE* 4. 3 (*GCS* 23. 148). Cf. Plotinus's use of 'fragrance' and 'light' in *Enn.* 5. 1, quoted in *PE* 11. 17 (*GCS* 43. 2. 38).

[88] *DE* 5. 1 (*GCS* 23. 211–12); Stead, *Divine Substance*, 231–2.

ority', which would be improper to the Son's divinity.[89] Rather, to be 'the Son, sole begotten of his will' is an assurance of the Son's divinity and intimacy with the Father; he is the only mediator.[90] However, the sense of will is not that of dynamic emanationism, as is clearly shown in the modification of the Wisdom metaphors. Nor is it solely an appropriate causality by analogy with creation, since the Son does not come from nothing. Instead, there is a sense of the deliberate appointment and direct derivation of a second divinity from the first, with an emphasis on separation and likeness. Thus, the Son is the living, essential image of the Father by the choice of the Father.

In Origen and the Middle Platonists the single essence of the highest god was commonly contrasted with the lower multiplicity of the second god, whose cosmological mediation required a multiple essence. Eusebius, however, does not draw this distinction between the essence of the Father and that of the Son, insisting instead that the Son is one, as the true image of the Father's essence:

> For the Father is one, so the Son is one, not many, and is the perfect one begotten, God from God, not several. . . . So one God has one Son, perfect and only begotten, but not many gods nor many sons of the Father. (*DE* 4. 3 (*GCS* 23. 152–3))

Eusebius's application of *μονάς* and *εἷς* to the Son's essence as the express image (Heb. 1: 3) of the Father is in direct contradiction to Origen and Numenius, who saw the multiplicity of the Son or second principle as necessary to his cosmological mediation.[91] Thus, although he stresses as strongly as Origen or Numenius the role of the Son as cosmological mediator between the transcendent essence of the Father and the lower weakness of the created order, he sees the Son's essence as not multiplied or dissipated as a result of this activity. Rather, the very likeness and uniqueness of the Son is why he is the one power of God (1 Cor. 1: 24), which, as a single force, pervades and orders all creation (Wisd. 7: 22).[92] The Son is clearly secondary as derivative

[89] *DE* 4. 3 (*GCS* 23. 153–4).

[90] *DE* 4. 5 (*GCS* 23. 157–8), 4. 2 (*GCS* 23. 152); *PE* 7. 15 (*GCS* 43. 1. 392).

[91] Compare *Jo.* 1. 22 (*SC* 120. 128) and Numenius, Fr. 11 and 16 (ed. des Places) with *DE* 4. 3 (*GCS* 23. 152) and 4. 5 (*GCS* 23. 155).

[92] *DE* 4. 5 (*GCS* 23. 155–6), 4. 6 (*GCS* 23. 158–9).

from the Father, which is what makes him able to be the mediator; yet his essential likeness to the Father is not compromised by his cosmological function.

Although the Son receives his being directly from the Father, his titles and essence are derivative, hence not identical. As in the case of Origen, Eusebius's strongest statements about the unity of the Father and the Son are drawn from the description of the Son as the image of God. Contrasting the *ἀγένητος*, hence unique essence of the Father with the begotten essence of the Son, Eusebius repeatedly underlines their unique relation rather than any shared nature. Thus, *εἰκών* and *μονογενής* are linked several times in Eusebius's discussions of the divinity of the Son. In his letter to Euphration of Balanea in 318, Eusebius used *εἰκών* as a way of understanding John 17: 3: the Son is the image of the true God, and as image is necessarily separate.[93] Image is thus a means of describing the distinct, if unique, status of the Son in relation to the unbegotten Father:

> Nor was he brought into being from the substance of the unbegotten by way of any passion or by division, nor was he eternally co-existent with the Father, for one is unbegotten and the other begotten and one is father and the other son, pre-existing since all agree a father precedes a son. So, the image of God is a kind of living image of the living God, but not such an image . . . which differs essentially, but one which contains the whole of its species, and is like it in its own essence to the Father. (*DE* 5. 1 (*GCS* 23. 213))

Because of the divinity of both, Eusebius declares that the separate, secondary hypostasis of the Son as Image does not violate monotheism, but makes the Son a distinct, living entity.[94]

Thus scriptural metaphor offers a way to understand the essential connection between *ἀγένητος* and *γένητος* divinity. As outlined above in regard to the Father, Eusebius denies that the essence of the Father can be shared by anything. He therefore supports the Arian claim, *ὁ ὢν τὸν μὴ ὄντα ἐγέννησε*, but emphasizes the qualifier that the Son as begotten is not like one of the creatures.[95] With regard to the use of will in generation, Eusebius is anxious to give the Son a separate and divine

[93] *Ur*. 3 (ed. Opitz, 6).

[94] *ET* 2. 20 (*GCS* 14. 129; cf. *ET* 2. 17 (*GCS* 14. 120), 3. 2 (*GCS* 14. 142–3).

[95] *DE* 5. 1 (*GCS* 23. 211); *ET* 1. 8 (*GCS* 14. 66), 1. 10 (*GCS* 14. 68).

hypostasis; he uses the terms *δεύτερος* or *γέννημα* to describe the secondary divinity of the Son.[96] The generation before time of the Son safeguards his divinity, and also separates him from other creatures.[97] In his letter reporting on the Council of Nicaea, Eusebius accepts direct generation, though not in any physical sense of shared essence, but in a way to underline the unique relationship of the Father and the Son. Thus Eusebius is close to the Athanasian understanding of the eternal and special generation of the Son from the Father, except that he denies any identity of essence between them. The Son can be from (*ἐκ*) the Father.[98]

Eusebius's intentions regarding the relation of essence between the Father and the Son may be further clarified by a comparison of his use of *αὐτο*-categories with that of Origen and Athanasius. In contrast to both, Eusebius underlines the derivation of all the Son's qualities while at the same time maintaining a likeness with the Father. Origen, on the basis of biblical exegesis, gave certain names exclusively to the Son (*αὐτοσοφία, αὐτοζώη*) or to the Father (*αὐτοθέος*); thus the Son was *θέος* by derivation, but *αὐτολόγος* in his essence and role.[99] Following Origen, Athanasius used *αὐτο*-categories to define the divinity of the Son, but claimed that the Son alone was the Wisdom or Word of the Father.[100] However, Eusebius bestows the various titles on both the Father and the Son, while reserving their origin to the Father alone; hence the Father is *αὐτοσοφία* or *αὐτολόγος*.[101] He describes the Son as the wisdom and power of God by the gift of the Father; the wise Father gives wisdom to the Son.[102] It is a sign of their essential unity that the Son is called Wisdom, Word, Power, or the Will of God; but he is not this on his own, but only from the Father. Thus in his own way Eusebius underlines both the derivation of the Son and his essential likeness to the Father. The emphasis on likeness of essence, so that both the Father and the Son are called *αὐτοαγαθόν* is also in direct con-

[96] *PE* 7. 15 (*GCS* 43. 1. 391); *HE* 1. 2. 5.

[97] *DE* 4. 3 (*GCS* 23. 154); *Ur.* 3 (ed. Opitz, 4–5).

[98] *Ur.* 22 (ed. Opitz, 45); *ET* 1. 8 (*GCS* 14. 66); see Stead, *Divine Substance*, 239–40.

[99] For use of *αὐτοσοφία* and *αὐτολόγος*, see Origen, *Cel.* 3. 41 (*SC* 136. 96); for *αὐτοθέος, θέος*, see idem, *Jo.* 2. 2 (*SC* 120. 216). See Gruber, *ΖΩΗ*, 104f.

[100] Athanasius, *Gent.* 46 (ed. Thomson, 130).

[101] *ET* 2. 14 (*GCS* 14. 115); *DE* 4. 2 (*GCS* 23. 151); *ET* 1. 8 (*GCS* 14. 66–7). See Arius, *Thalia*, in *Ar.* 1. 5 (ed. Bright, 5).

[102] *DE* 4. 2 (*GCS* 23. 152).

trast to the hierarchical structure of Middle Platonism and probably of Origen. Eusebius claims that the Son receives the titles to be the co-worker of the Father's will.[103] Thus he describes a stronger unity and likeness in terms of shared qualities between the Father and the Son than do either Origen or the Middle Platonists; yet, by comprison with Athanasius, he underlines the derivation or appointment of the Son to a sharper degree, even in his essential likeness.

Not surprisingly, participation is also modified by Eusebius, by comparison with Origen and the Middle Platonists. He generally uses *μετέχω* and *μετουσία* to express the unique derivation and relation of the Father and Son, but with a limited sense of an essential connection.[104] By his use of Psalm 44: 7–8 and Isaiah 61: 1–2, Eusebius, in line with his interest in derivation rather than essential communication, associates participation with prophetic anointing.[105] Thus in the *Ecclesiastical History*, Eusebius writes: '*ἀλλ' αὐτὸ δὴ σνεύματι θείῳ τὸ θεοτρεπές, μετοχῇ τῆς ἀγεννήτου καὶ πατρικῆς θεότητος ἀπειλήφει.*' In the *Demonstration* 4. 15, *μετουσία* is again associated with Psalm 44: 7: '*μόνον ἐξ αὐτοῦ γεννηθέντα θεὸν λόγον, μετοχῇ τε τοῦ γεννήσαντος ἀγεννήτου καὶ πρώτου.*' Throughout this chapter Eusebius uses biblical accounts of anointing to describe the unique relation of the Son who receives power and divinity from the unbegotten Father.[106] By this anointing—that is, this direct relation to divine power and essence—Christ is not a creature; but he is divine, for he is the first and the only one to be anointed with 'full' oil. Eusebius's use of these passages to describe the begetting of the pre-existent Son may be a deliberate attempt to avoid any adoptionist interpretation of the prophecies. However, the description also reflects Eusebius's usual pattern of describing the generation of the Son by the Father through scriptural images. Eusebius's use of *μετέχω* suggests directly derived relation, rather than substantial communion.[107] In a few passages

[103] *PE* 7. 15 (*GCS* 43. 1. 391–2); *Theo.* 1. 23 (*GCS* 11. 2. 46f.).

[104] Eusebius is usually assumed to follow Origen; see Gregg and Groh, *Early Arianism*, 127 n. 136; Williams, *Arius*, 222.

[105] Weber, Ἀρχή, 59–67; Gregg and Groh, *Early Arianism*, 199.

[106] *DE* 4. 15 (*GCS* 23. 175); cf. *DE* 4. 15 (*GCS* 23. 182): one god anoints, another receives.

[107] *DE* 9. 10 (*GCS* 23. 427); cf. *ET* 1. 2 (*GCS* 14. 63); 2. 17 (*GCS* 14. 121), 3. 18 (*GCS* 14. 179).

Eusebius resembles Origen, describing the Son as the one who receives a secondary divinity by participation.[108] But Eusebius's use of participation language in conjunction with the scriptural image of anointing and in relation to the unbegotten nature of the Father suggests that he intends to underline the unique, essentially derivative relation rather than a shared essence between the Father and the Son.

This explicit separation of the Father and the Son in Eusebius's theology has sometimes been seen as deriving from a preoccupation with the philosophical problem of transcendence, so that his concern with cosmological mediation leads him to a more radical subordination. However, as shown in these comparisons with Origen and the Middle Platonists, Eusebius proposed a more essential likeness between the Father and the Son, even though he emphasized the primacy of the Father's essence as *ἀγένητος*. The Son's cosmological activity does not affect his essence in regard to the Father; in fact, to be the authentic intermediary, the Son must share essential qualities with the Father to a greater extent than he is represented as doing in Origen.

Eusebius's synthesis of Platonic ontology and biblical images of agency is also apparent in his defence of Asterius against Marcellus. The Son is not an impersonal wisdom, but rather the living Son of God, the unique mediator and leader of worship.[109] Fearing modalism and adoptionism in Marcellus, Eusebius describes the necessity of the Son's separate *ὑπόστασις* which is proved by his actions and words in Scripture:

> But we say that one is unbegotten and without beginning while the other is begotten and has the Father as a beginning. Therefore, the Son himself teaches that his Father is God when he says to them, 'I am going up to my Father and your Father, to my God and your God.' . . . Thus, one God is proclaimed by the Church. . . . He [the Son] honours, adores and glorifies his own Father as God, . . . and he is recorded as thanking him, praying to him and as being obedient unto death. He confesses that he lives 'because of the Father' and that he can do nothing without the Father, that he does not his will, but that of his Father. (*ET* 2. 7 (*GCS* 14. 104))

Here Eusebius directly links the secondary being and status of the Son with the scriptural accounts of the prayers and obedience

108 Compare *DE* 5. 4 (*GCS* 23. 226) with Origen, *Jo.* 2. 17–18 (*SC* 120. 284–8).
109 *ET* 2. 17 (*GCS* 14. 120), 2. 23 (*GCS* 14. 133), 3. 2 (*GCS* 14. 139).

of Jesus. This passage reflects more than economic subordinationism or cosmological mediation in the profound link between the cosmological identity and the historical activity of Jesus. Rather than merely a philosophical intermediary, the cosmic agency of the Son in Eusebius's theology necessarily entails the obedient activity of the biblical Jesus; he is the one agent in two lives. His identity distinct from that of the Father is necessary not only on account of ontological concerns, but also as a way of making sense of the scriptural account of his independent action. Cosmologically this identity remains constant; in the 'two lives' the Logos both reveals the Father and embodies a model of human perfection, holding together the cosmic and historical action by the consistency of the agency of the single subject.

The combination of the agency of the Platonic Logos with scriptural references to the obedience of the Son thus pervades Eusebius's work. In *DE* 4. 4 the Son is subject to the headship of God (1 Cor. 11: 3) to be Choregus of Light and Demiurge. Elsewhere Eusebius uses Psalm 2: 7 to show the Physician of the universe:

> And he is called sun of righteousness (Mal. 4: 2) and the true light (John 1: 9), carrying out and co-operating in his Father's commands, so he is also called minister of the Father and creator. For he alone according to rank knew how to serve God, and stands between the unbegotten God and the things after Him begotten, and has received the care of the universe and is priest to the Father on behalf of all who are obedient (Heb. 6: 20). . . . He is called eternal high priest, and also the anointed. (*De* 4. 10 (*GCS* 23. 167))

In this passage Eusebius mixes not only Platonic titles of cosmological mediation, but also descriptions of Christ from the New Testament. As discussed above, Origen usually distinguished between certain titles; for example, he used the high priest image of Hebrews for the incarnate Logos, and described Jesus as midway between the unbegotten God and begotten creation. Eusebius makes no such distinction, applying all language about the cosmological and historical tasks to one subject, the Son.

The difficulty with Eusebius's Christology, therefore, seems to lie as much in a problem of scriptural interpretation as in philosophical cosmology. Undoubtedly the hierarchical structure of Middle Platonic divinity lent itself to his Christian Logos theology. Yet applying a hermeneutic of John 17: 3 or 1 Corinthians 15: 28 to 1 Corinthians 1: 24 or John 1: 1—as Eusebius clearly does—would produce a similar result. That is to say, if a cosmology were constructed on the basis of the biblical record of the relationship of the Father and the Son, an ontological hierarchy and separation would naturally result, a hierarchy and separation more striking than in contemporary Platonism because of the personal relational language of obedience and worship. To a degree, Eusebius's scriptural language at times worked against his efforts to describe the essential intimacy or unity of the Father and the Son by means of derivation of qualities or image. Eusebius's assertion of monotheism, which he supports by reference to Scripture as well as to philosophy, makes it impossible to maintain a coherent account of the Son's divinity and soteriological role apart from a hierarchy of divinity.

Focusing on the soteriological role of the Logos as the power of God, Eusebius therefore emphasizes his continuous character, as agent in both pre-existence and incarnation. Like his contemporaries, he tries to make sense of the incarnate experiences of the divine Son without consistent reference to a human soul, and encounters difficulties in the interpretation of emotion. In his earlier exegetical work on the Psalms, Eusebius occasionally refers to a human soul; but in general he sees the Logos as the active force within the incarnate Christ.[110] He argues at length that the Word of God was present to human experience, but was not hampered by the body or corrupted by it:

> Being the Word of God, he remained by nature without body or dwelling in the flesh, and went through the whole economy with divine power and in ways unknown to us, sharing what belonged to him, but not receiving what belonged to others. (*DE* 7. 1 (*GCS* 23. 302))

By his coming in a human instrument (*ὄργανον*), the Logos showed empathy with the body, thereby allowing humans to see God, respond to something like themselves, and so receive the power of God for deification.[111] This emphasis on the revelation or theophany of the Logos seemingly buffered from human

[110] G. C. Stead, 'The Scriptures and the Soul of Christ in Athanasius', *VC* 36 (1982), 233; Hanson, *Christian Search*, 53–5; Williams, *Arius*, 174.

[111] *Theo*. 3. 39 (*GCS* 11. 2. 140–1); *DE* 4. 13–14 (*GCS* 23. 172–3), 3. 6 (*GCS* 23. 131).

experience has led to an assessment of his soteriological role as external: that is, evidential or demonstrative, rather than constituting an authentic, transforming union between divine and human nature.[112] However, the relationship between the Logos and his body echoes the 'double nature' of every human person in Eusebius's theological model: material existence is a created, temporary state which can be transformed by the proper exercise of the soul. The incarnate Logos in his 'double nature' thus represents both the full power of divinity and the model of ideal virtue: that is, the absolute control of the body by the Logos.

Thus, Eusebius typically treats the life of Jesus as a testimony to the reality of divine power in history. The reason for Jesus's unique life and acts was his source of virtue, the Father; in all his acts and teaching, he referred to the Father's power.[113] If, on the one hand, Eusebius emphasizes the descent of the Logos as necessary to reveal divine nature, on the other, he wishes to examine the life of Jesus, whose unique power and piety prove that he was indeed the Son of God. From his earliest to his final works, Eusebius celebrates the unique, hence divine power of Jesus in the miracles. By these, Jesus awakened the world to the active presence of God's mercy and power in material life, and thereby established true morality and worship.[114] Indeed, the very name 'Jesus' means salvation, and points to his divinity and to his Father:

> So then, his 'brothers' at first praised him only as a remarkable man because of his miracles, believing him to be like one of the prophets; but afterwards when they saw his wonderful miracles, such as how he destroyed the enemy and the avenger and death . . . they believed him to be God and worshipped him. (*DE* 8. 1(*GCS* 23. 363))

Those who receive the teaching of Jesus and confess his divinity show the power of God in their lives and preaching; the evidence of divine power in the spread and triumph of the church and in continuing miracles show the truth of Jesus's divine origin.[115]

[112] See Young, *Nicaea to Chalcedon*, 69, *re* the evidential value of the death of Christ, as opposed to the transformative soteriology of Athanasius. Cf. J. Roldanus, *Le Christ et l'homme dans la théologie d'Athanase d'Alexandrie* (Leiden, 1968), 19.

[113] On the importance of this apologetic theme, see Wallace-Hadrill, *Eusebius of Caesarea*, 105–8.

[114] *DE* 9. 13 (*GCS* 23. 432–3). [115] *DE* 1. 10 (*GCS* 23. 45), 3. 2 (*GCS* 23. 107).

Transforming works, not only teaching, were essential in Christology to prove and establish the validity of the divine presence in Jesus.

This revelation and communication of divine power took place through the human life of the divine Logos. Eusebius examined the Hebrew Scriptures for confirmation of the historical details of Jesus's life to show that the incarnation of the Son was foretold. Here he used *πρόσωπον* as an exegetical device for applying the words of the psalmist or prophet to the person of Christ.[116] He also used it as a means of presenting the Incarnation itself: 'He promises that the God of God who dwells in Christ will manifest himself in the person of Christ.'[117] On occasion, in his own writings, Eusebius describes the humanity of Christ as a *πρόσωπον*, in the sense of the subject of the action. Thus 'Christ' was perceived as and behaved like an ordinary person, even if, eventually, he was discovered to be the Logos. He went out into the desert to be tempted as a human being; yet the purpose of his temptations was divine—namely, to confront the demons with his divine power.[118] Hence Jesus was the special child prophesied as Emmanuel: '[He] is to be endowed with more power than the typical human being, choosing good before knowing evil, and this not in maturity, but as a child' (*DE* 7. 1 (*GCS* 23. 305)).

Through the human *πρόσωπον* the Logos exemplified the perfect obedience of ideal human nature. In *DE* 10 Eusebius uses a remarkable first-person narrative to explore the fear and obedience of the Son on the cross. Working from Psalm 21: 2–32, he reconstructs the thoughts of Jesus:

> And in this hope [of resurrection] I trust you, my God and Lord and Father, and this is not the first time I put my hope in you. . . . For while still carried in the chamber of her that bore me, I saw you, my God, being separate and untroubled although joined to the flesh, being incorporeal and free from all ties. . . . Holding such memories in my mind, and through all things setting my God and my Father before my eyes, likewise in this time of supreme suffering I should do the same, when in submission [*ἑκὼν*] to you, my Father, by my own choice and free will [*αὐθαίρετος*] I became a worm and no man. (*DE* 10. 8 (*GCS* 23. 481–2))

[116] *DE* 6. 4 (*GCS* 23. 255), 5. 13 (*GCS* 23. 236).

[117] *DE* 9. 7 (*GCS* 23. 418).

[118] *DE* 4. 10 (*GCS* 23. 167).

In this curious description Eusebius attempts to affirm the reality of the divinity of Christ, as well as his obedience to the Father. In light of his anthropology, the account is not docetic. The power of the Logos is such as to allow Eusebius to speak of Christ's body as 'free' in opposition to other bodies which are 'slave'; thus the power of the 'soul' is in complete control of the flesh. Since the Christian ideal is control of the body through empowerment of the soul by the Logos, as seen in the lives of the martyrs and ascetics, the dominance of Logos over body in Eusebius's Christology thus represents a moral ideal.

In general Eusebius is more concerned about docetism than about problems of authentic obedience within the psychology of the incarnate Logos. As seen above, *πρόσωπον* is used to affirm the reality of suffering or obedience as appropriate to the incarnate life:

> But observe how all this is said, in the person [*πρόσωπον*] of him who was carried in a mother's womb, and born of a mother, whom we call the Lamb of God. For the words about passion as those about the incarnate birth apply to him. . . . This our Saviour and Lord recounts, not as being without flesh and body nor as he is regarded as the word of God and divine, but as the one who prayed to the Father. (*DE* 10. 8 (*GCS* 23. 483))

In the activity of the incarnate Son is seen the model for human imitation, although this action is consistent with the character and relation of the Logos to the Father as divine agent. Thus, in Eusebius, both birth and death reflect the self-emptying love of the Son in accordance with the will of the Father.[119] In his account of the Passion, divine self-giving and human obedience are intertwined:

> Then, his strong one abandoned him, desiring him to go to death and the death on the cross and to be made known as the ransom and sacrifice for the whole world and to be the purification of the life of them that believe in him. And he, since he understood immediately his Father's will, and discerned why he was abandoned by the Father, humbled himself even more, and embraced death for us with all eagerness [*προθύμος*]. . . . And the Father abandoned him for another reason, namely that the love of Christ himself for humanity might be set forth.

[119] *DE* 10. 8 (*GCS* 23. 485), 9. 7 (*GCS* 23. 419).

For no one had power over his soul, but he gave it willingly for humanity. (*DE* 10. 8 (*GCS* 23. 477–8))

Unlike Origen's heroic Son, Eusebius's Christ is not ignorant of the purpose of his death, but discerns and accepts the Father's will in voluntary humility. Since the power of God has filled and guided his entire sojourn, the Son continues to rely on the will of God in spite of his suffering. In contrast to Athanasius, who saw the inner Logos as conquering human fear in Jesus, Eusebius focuses on the obedience of Christ to the Father, to prove 'the glory and independence of his own choice'. His obedience in human form conveys a consistent character to his relation to the will of the Father. The will of the Son was therefore divine, but whether pre-existent or incarnate, it had the same obedient relation to God. The theological point for Eusebius is the power one receives by submission to divine will. Thus, when Christ confronted the demons, they drew back in surprise, since they found not a human soul, but the Logos.[120]

Using *πρόσωπον* as a literary device to present the one subject in two lives, Eusebius attributes both power and obedience to the saving work of the Son of God as both Logos and Jesus. To deliver humanity from ignorance of God and restore knowledge of God, the divine Son became incarnate.[121] This agency of the Son expresses his obedience to the Father; he was sent to do the Father's wishes.[122] The incarnate Logos, as the living image of God, overturned idols: but he was also the priest who was the example of piety toward the Father.[123] Thus, in the work of Jesus, one sees 'in truth God and the Word; see how he existed by an interpreter with men, in the example of his sufferings . . . healed disease and infirmity, and how ready he was, in whom there was no sin, to do good works.'[124]

Eusebius therefore uses a sort of double exegesis to determine which mode or life a particular text portrays of the subject. The human *πρόσωπον* of Christ properly narrates the Passion, just as the Logos speaks to the prophets; yet it is the same Son of the Father. While this sort of exegesis is hardly systematic in

[120] *Theo.* 3. 44 (*GCS* 11. 2. 148).

[121] *DE* 3. 4 (*GCS* 23. 106), 6. 16 (*GCS* 23. 272).

[122] *Theo.* 3. 39 (*GCS* 11. 2. 140–2); *HE* 1. 2. 7; *ET* 2. 6–7 (*GCS* 14. 103–4); *HE* 10. 4. 67.

[123] *Theo.* 3. 40 (*GCS* 11. 2. 145–6).

[124] *DE* 10. 8 (*GCS* 23. 481).

Eusebius, it allows him at various points to present the humanity of Christ as more than a temporary body or instrument of the Logos. Rather than attributing the fear of Jesus to human weakness, Eusebius portrays the words on the cross as an example of obedience to divine will for human faith. Thus, the Son who became incarnate in accordance with the Father's will is recognizable in Christ who submits to death in obedience to the Father's will and out of love for humanity. This allows a consistency of behaviour essential to the pedagogical character of his Christology in which the incarnate Logos truly reveals both divine nature and ideal humanity.

Salvation thus rests on the revelation and communication of divine power brought about by the incarnation and actions of the Son. As incarnate, his teaching, death, and resurrection demonstrate the divine power over death and its defeat of the power of the demons; his body was also a ransom which bore the sins of humanity, purified the flesh, and redeemed them into incorruptibility. By his incarnate teaching and healing, the Logos brought divine power and transformation:

> And he outraged these (demons) by his works and words, while he healed and cured the whole human race with the gentle and kind medicines of his words and tonic of his teaching. He freed them from all sorts of sickness and suffering not only of the body but also of the soul. . . . In addition to all this he threw open the gates of heavenly life and by his holy teaching, to all nations, and so extended his hand not only to the sick . . . but also to save the half-dead . . . and the dead and buried. (*DE* 4. 10 (*GCS* 23. 167))

Just as the acts of the church reveal and communicate the will of God to the broader society, so the acts of Christ reveal the nature of God and have established the means of redemption by the transformation of human nature. Theophany does not consist in mere appearance, but in the authentic and decisive intersection of divine power and creation. Although Eusebius emphasizes the teaching of the Logos, it is hardly an external or merely moral instruction, but rather constitutes access to transforming power to heal both soul and body.

Cosmology in Eusebius is thus the universal expression of the historical event of Jesus's teachings. From the perspective of later orthodoxy, one may reduce Eusebius's soteriology to the external instruction of a cosmic agent, as opposed to the transformation of human life by divine incarnation. But such an evaluation ignores the interweaving of cosmic and biblical language, as well as the personal and social transformations achieved through the power of Christ. The incarnation of the Logos communicates the power of God which ensures individual progress. Against fatalism, social chaos and reform, and increasing philosophical transcendence, the Christological acts of God, for Eusebius, brought reassurance and evidence of the accessible purpose and power of divinity in a changing world. The courage of the martyrs and the growth of the church through the teachings of Jesus reflect the individual and social harmony which will eventually come to echo the cosmic unity led by the Logos; the social and material world will thus be transformed in the same fashion as individual bodies through the power of God. Equally important, as cosmic agent and obedient Son, the Logos embodies both divine will and human perfection for fallen humanity. In this way Eusebius seeks to present authentic incarnation and moral example in his Christology. As an apologist, Eusebius thus presents an optimistic and universal account of divine power and human progress founded on the historical evidence of Jesus's life and the triumph of the church.

4

ATHANASIUS TRANSCENDENCE AND TRANSFORMATION

> For he suffered to prepare freedom from suffering for those who suffer in him; he descended that he might raise us up; he took on him the trial of being born, that we might love him who is unborn; he went down to corruption, that corruption might put on immortality.
>
> (Athanasius, Festal letter)

'Of such a one the times had the greatest need': Sozomen's evaluation of Athanasius has largely persisted in the history of fourth-century Christianity.[1] His tempestuous career as ascetic bishop, religious polemicist, and ecclesiastical politician testifies to his doctrinal integrity and personal tenacity; yet his leading role in the Arian controversy has made his theology difficult to evaluate historically. In contrast to the philosophical and apologetic concerns of Origen and Eusebius, Athanasius's passionate defence of the incarnation of Christ seems to offer a purely religious cosmology. This 'soteriological' or 'Christocentric' emphasis is a commonplace in scholarly literature: the incarnate Christ was not the culmination of pedagogical cosmology or history, but rather the key to the salvation of all existence.[2] However, as Paul Veyne has pointed out, 'A key is not an explanation. While an explana-

[1] *HE* 2. 17. Studies of Athanasius and Arianism continue to emphasize his passionate defence of orthodoxy; see C. Kannengiesser, *Athanase d'Alexandrie* (Paris, 1983); cf. idem, 'The Athanasian Decade 1974–1984', *TS* 46 (1985), 524–41; Gregg and Groh, *Early Arianism*, 170f.; Willams, *Arius*, 233f.; Ricken, 'Nikaia'.

[2] C. Kannengiesser, 'Athanasius of Alexandria and the Foundation of Traditional Christology', *TS* 34 (1973), 112; idem, *Athanase*, 113–254; Ricken, 'Nikaia', 339; idem, 'Rezeption', 349–52; de Vogel, 'Platonism and Christianity', 50; Weber, Ἀρχή, 177–8.

tion accounts for a phenomenon, a key makes us forget the riddle . . . ; the solution is the whole point.'[3]

As the key to later orthodoxy, Athanasius's Christology has often obscured the historical and theological riddles of his cosmology. This study will follow the work of E. P. Meijering, C. Kannengiesser, G. C. Stead, R. Gregg, and D. Groh in attempting to expose and explore the philosophical and ascetic assumptions which underlie his familiar and dramatic defence of the Incarnation. Working from a definition of a theological model, his work is no less cosmological or more soteriological than that of earlier authors; rather, it is shaped by the passionate and narrow clarity of a fourth-century bishop.

ATHANASIUS AS BISHOP

Throughout his career of over four decades as bishop of Alexandria, Athanasius combined administrative, political, and theological skills to bring unity and to ensure Nicene orthodoxy in the Eastern church. Not only the Christian Platonic tradition, but regional conflicts, civil persecution, and Western ties shaped the Alexandrian episcopacy in the fourth century. The divisive rigorism of the Meletian controversy, the uncompromising alternative vision of the ascetic movement, and emerging Coptic Christianity offered a complex challenge to Alexandrian theology, as well as to ecclesiology. With increasing imperial support of Christianity, the bishop became a symbol not only of the spiritual community, but of the diverse social elements within it; bishops were the pivot between government and those who were largely unseen in Late Roman society: the poor and women.[4] Arguing from inherited traditions, liturgical practice, and a growing vocabulary of asceticism and orthodoxy, bishops therefore became architects of both social and doctrinal unity. Over the course of the fourth century, the security of ecclesiastical position lay increasingly in doctrinal positions or the ability to control

[3] Veyne, *Did the Greeks Believe their Myths?*, tr. P. Wissing (Chicago, 1988), 30.

[4] W. H. C. Frend, 'Athanasius as an Egyptian Christian Leader in the Fourth Century', in idem, *Religion Popular and Unpopular in the Early Christian Centuries* (London, 1976), 20–37; H. Chadwick, *The Role of the Christian Bishop in Ancient Society* (Berkeley, Calif., 1980), 5f.

theological controversy with rivals. Imperial patronage, the decades of the Arian controversy, and increasing conversions altered the role of bishop from local teacher to cosmopolitan representative of ecclesiastical and often secular authority. Athanasius embodied both the old ideals of a separate Christian identity and the new social realities of a public, imperial church.

Coming to maturity in the midst of the Meletian and Arian crises, Athanasius was by necessity committed to church unity and orthodoxy. Born in Alexandria around 295, he was educated in philosophy and rhetoric; he seems to have had ties with the ascetic community, as well as with the household of Bishop Alexander, before his election to the episcopate in 328.[5] From his early apologetic writings to his mature theological works against the Arians, he maintained without question certain theological preferences and assumptions.[6] Undoubtedly, his stormy career pressed his writing, if not his thinking, into absolute categories; yet his theological confidence seems to have been rooted in the new Constantinian era of Christian social power, ascetic renewal, and doctrinal definition. By temperament and perspective, he was far from the passionate intellectual openness of Origen or the weathered apologetic optimism of Eusebius. Instead, he heralded a new age of ascetic bishop characterized by 'idealistic rigidity'.[7]

To ascetics, emperors, and Copts, Athanasius as bishop embodied a union between orthodoxy, community, and holiness. Securing his authority beyond the city of Alexandria, he integrated the ascetic, Coptic perspective into earlier Alexandrian teaching by his travel, support, and friendship with monks. In turn the growing prestige of the ascetics helped to buttress his own ecclesiastical and theological power; his *Life of Antony* portrayed the desirable unity between desert life and the urban church.[8] To the people in the villages, the bishop became a power-

[5] Rufinus, *HE* 1. 14. On his early, obscure background, see M. Tetz, 'Zur Biographie des Athanasius von Alexandrien', *Zeitschrift für Kirchengeschichte*, 90 (1979), 158–92, and Kannengiesser, 'Athanasian Decade', 524–8.

[6] On the largely accepted dating of *De Incarnatione* in the 330s, see most recently M. Slusser, 'Athanasius, *Contra Gentes*, and *De Incarnatione*', *JTS*, NS 37 (1986), 114–17. See also the suggestion of his authorship of *Ἑνὸς σώματος* by G. C. Stead, 'Athanasius' Earliest Written Work', *JTS*, NS 39 (1988), 76–91.

[7] This phrase is taken from C. Kannengiesser, 'Athanasius of Alexandria vs. Arius', in Pearson and Goehring (eds.), *Egyptian Christianity*, 205.

[8] B. Brennan, 'Athanasius' *Vita Antonii*', *VC* 39 (1985), 209–27; Gregg and Groh, *Early Arianism*, 131–59; Frend, 'Athanasius'. On questions of authorship

ful and stable patron in an age in which Christianity was gradually replacing social status as a factor in local politics.[9] He judged emperors and officials by their doctrinal choices rather than by any blind loyalty to the state; Athanasius's view of Constantius changed according to the vicissitudes of church councils.[10] His rhetorical skill, personal tenacity, and compelling witness to the Nicene faith gained him an international following and increasing support. The model he established for Alexandrian bishops was one of uncompromising orthodoxy and political astuteness.

Theologically and ecclesiastically, the Arian controversy was the whetstone for Athanasius's mature exposition of Christianity. Recent studies remain divided on his role at Nicaea and in the earliest years of the debate; but clearly he became a leading defender of Nicene orthodoxy against its theological opponents in the East and the West.[11] This polemical context creates several problems for understanding Athanasius's theology. First, most of it was presented in a highly rhetorical, uneven style, which at times does justice neither to his opponents nor to his own intentions.[12] Second, in spite of current revisions of the Arian controversy itself, Athanasius has retained a remarkably consistent character as a traditional teacher of orthodoxy. Gregg and Groh emphasize his 'essentialist' view of the world, heightened by the Egyptian ascetic context in contrast to Arius's 'voluntarist' view.[13] Williams and Kannengiesser portray him as a 'Catholic traditionalist' prodded to new reflection by controversy, rather than an academic and a philosopher like Arius. They follow Ricken and Vogel in stressing Athanasius's soteriological, hence

of the *Life of Antony*, see A. Louth, 'St. Athanasius and the Greek *Life of Antony*', *JTS* 39 (1988), 504–9.

[9] E. A. Judge and S. R. Pickering, 'Papyrus Documentation of Church and Community in Egypt to the Mid-Fourth Century', *Jahrbuch für Antike und Christentum*, 20 (1977), 69.

[10] Compare his attitudes in *Apologia ad Constantius*, *Historia Arianorum*, and *De Synodis*; cf. L. Barnard, 'Athanase, Constantin, et Constance', in Kannengiesser (ed.), *Politique et théologie*, 128f.

[11] Compare Kannengiesser, 'Athanasius of Alexandria', 107, and Stead, 'Earliest Work'.

[12] G. C. Stead, 'Rhetorical Method in Athanasius', *VC* 30 (1976), 121–37; In a review, he commented, 'I also think that sometimes, although the answer he presents to a problem as he conceives it is wholly convincing, the problem itself is misconceived' (*JTS* 36 (1985), 226).

[13] Gregg and Groh, *Early Arianism*, 170f.

essentially Christian concerns: *creatio ex nihilo*, the incarnation of the divine Son, and the divinization of embodied humanity.[14]

Athanasius was clearly not primarily a creative theologian, but an episcopal, ascetic teacher in a new age of Christian identity. His straightforward defence of Nicene orthodoxy has the ring of an *abba* as well as a *papas* about it; he is not an apologist in dialogue with a majority culture, but an authority in conflict with his own tradition. Although he drew on the apologetics of Eusebius, he was not overly concerned with the congruence of Christianity and Graeco-Roman culture or even with the progress of salvation history.[15] Instead he focused on the exclusive identity of Christians apart from the world; salvation lay within the Church, which had received the revelation of God. History, therefore, was not a school for souls, but rather the chaotic backdrop of human estrangement which the determined ascetic sought to avoid. Hence Christian belief should not resonate with echoes of the world; rather, belief in Christ exposes worldly folly, idolatry, and magic.[16] Athanasius's episcopal and ascetical writings are therefore critical to an understanding of the framework of his theological reflection. Earlier apologetic theological arguments acquire a different weight and focus in the context of fourth-century struggles for ecclesiastical unity and doctrinal consensus.

Sharpened by asceticism and episcopal authority, Athanasius's theological method therefore rested squarely on the revealed word of God as found in the Christian community. Because of sin, divine knowledge was beyond the capability of creatures, and it was only through incarnate divine nature that God was

[14] Kannengiesser, 'Athanasius vs. Arius', 213–14; idem, *Athanase*, 113; Williams, *Arius*, 233f.

[15] On his tone as a 'winner', see Rist, 'Basil's Neoplatonism', 50–4; E. P. Meijering, *Orthodoxy and Platonism in Athanasius* (Leiden, 1968); Meijering described *Against the Pagans* as composed for a war already won (idem (ed.), *Contra Gentes. Introduction, Translation, and Commentary* (Leiden, 1984), 154–5). For a summary of the scholarship regarding the possible relation of Eusebius's early apologetic works to the writings of Athanasius, see M.-J. Rondeau, 'Une nouvelle preuve d'influence littéraire d'Eusèbe de Césarée sur Athanase', *Recherches de science religieuse*, 56 (1968), 385–434. Meijering discounts decisive dependence on *PE* because of the commonplace nature of the parallels (*Contra Gentes*, 3).

[16] On the contrast between world and desert, see Brown, *Body and Society*, 216; cf. *Inc.* 46 (*SC* 199. 434–6), 50 (*SC* 199. 446–8), 57 (*SC* 199. 466–7); *Ar.* 2. 69 (ed. Bright, 139), 2. 72 (ed. Bright, 142–3).

known.[17] This epistemological assumption of the gap between divine and human nature, shared in more radical and distinctive form by Arius, increasingly shaped his reflection about Christ in a fundamental fashion: the eternal and incarnate son, who was the essential image of the Father, was the only guide able to reveal and embody divine truth for fallen humanity.[18] Although in his earlier works he offered traditional apologetic arguments for the knowledge of God, he later came to see the full, substantial divinity of Christ in the Incarnation as essential to the security of divine revelation, and hence to the authenticity of the community. Since faith in the incarnate Christ was the only way to the Father, 'heretics' such as Manichees or Arians had only 'mythology'.[19] Fables and speculation regarding human experience were quite distinct from the revealed knowledge of God in the Church. One could not look ultimately to creation to reveal God; without the Word, the instability and terror of the world revealed only human separation from God.[20]

Theological certainty thus rested on Scripture, which had absolute authority as the revealed divine Word. Here God has revealed the language appropriate to divinity.[21] One can speak only negatively with assurance because of divine transcendence and incomprehensibility; yet what is revealed is absolute, for essences precede words, and the terms applied to God in Scripture reveal the essential divine nature.[22] Of course Athanasius realized the difficulties of scriptural interpretation, which required that the ambiguity of Scripture be steadied by tradition. He referred often to the perplexity of the believer, but maintained that faith delivered one from destructive doubt or specious thought.[23] By having confidence in God's absolute

[17] *Ar.* 1. 2 (ed. Bright, 2), 1. 23 (ed. Bright, 24–5); cf. J. B. Walker, 'Covenance épistémologique de l'"Homoousion" dans la théologie d'Athanase', in Kannengiesser (ed.), *Politique et théologie*, 255–75.

[18] *Ar.* 1. 34 (ed. Bright, 35), 1. 36 (ed. Bright, 37), 3. 5 (ed. Bright, 158).

[19] *Ar.* 2. 43 (ed. Bright, 112f.); *His. Ar.* 66 (*PG* 25. 772).

[20] Compare *Gent.* 44 (ed. Thomson, 120) with *Inc.* 12–14 (*SC* 199. 306–18), 43 (*SC* 199. 420), 47 (*SC* 199. 438). Athanasius is not wholly consistent; see *Ar.* 2. 81–2 (ed. Bright, 151–2) and *Inc.* 15 (*SC* 199. 318f.).

[21] *Ar.* 1. 52 (ed. Bright, 54–5); *Decr.* 12 (*PG* 25. 445); *Ep. Serap.* 2. 1 (*PG* 26. 609); cf. H. J. Sieben, 'Herméneutique de l'exégèse dogmatique d'Athanase', in Kannengiesser (ed.), *Politique et théologie*, 195–214.

[22] *Ar.* 2. 3 (ed. Bright, 70–1).

[23] *Ar.* 2. 36 (ed. Bright, 105); *Ep. Serap.* 1. 17 (*PG* 26. 571).

revelation in Christ, Christians had a complete religion.[24] Faith in the divinity of Christ allowed one to understand both Scripture and tradition correctly.[25] Both aspects were necessary for correct belief, since error may result from either faithless reading of Scripture or from a faith that lacks scriptural anchoring. As one considered the intentions of earlier writers with a view to grasping their true meaning, one needed to weigh the entire context (*σκόπος*) of Scripture: namely, God's economy of creation and redemption.[26] The heart of scriptural interpretation, then, was faith: that is, a theologically correct conception of God's activity in Christ. Equally important, theology supported by the community followed faith, and faith followed God's revelation in Christ and in Scripture. In Scripture, God gave the correct terms for theological understanding, which were interpreted correctly only within the church. Thus, for Athanasius, scriptural interpretation rested on a 'skopos ecclesiastique'.[27]

In part owing to the bitter controversies concerning theological language and incarnation in the fourth century, Athanasius narrowed earlier apologetic epistemology by his focus on the sole revelation of God in the incarnate Christ. Argued against his later opponents, the essential relationship of the Son in preexistence and incarnation became the linchpin of Christian truth claims and authentic community. This perspective has sometimes been attributed to a Hebraic cast of mind, which rejected a Greek conception of cosmological continuity between God and humanity and took seriously the gulf between created and uncreated; Athanasius was more faithful to the New Testament kerygma of God in Christ or Christian catechesis than to the cosmological concerns of Greek Logos theology.[28] However, this epistemological focus on the incarnate Word also has a curious parallel with developments in contemporary religious thought.

[24] *Ar.* 1. 18 (ed. Bright, 19).

[25] *Ar.* 1. 9 (ed. Bright, 9–10), 3. 58 (ed. Bright, 211); *Ep. Serap.* 1. 17 (*PG* 26. 571).

[26] *Ar.* 1. 55 (ed. Bright, 57), 3. 29 (ed. Bright, 184); cf. Sieben, 'Herméneutique', 210.

[27] Sieben, 'Herméneutique', 213. See *Ar.* 1. 23 (ed. Bright, 24), 2. 43 (ed. Bright, 112). Athanasius contrasted individual (*idios*) and communal (*ecclesia*) interpretation; see *Ar.* 1. 37 (ed. Bright, 38) and 3. 10 (ed. Bright, 163).

[28] Ricken, 'Nikaia', 340; de Vogel, 'Platonism and Christianity', 50–1; cf. Williams on Barth and Athanasius, *Arius*, 238.

With the development of the idea of increased transcendence of the first principle, later Platonists, from Albinus to Plotinus, had rejected any use of analogy in attempting to understand God, relying on negative theology and self-communication through emanation.[29] The Manichees, in Egypt, emphasized the transcendence of God, and celebrated the decisive revelation through Christ; within their dualistic world-view this docetic Christ gave a personal, liberating knowledge to the elect.[30] Within the mainstream Christian community, Arius affirmed the revelation of the transcendent God through the agency of the created, but incarnate Christ; theological arguments later in the fourth century continued to focus on the proper epistemological basis for Christian belief.[31] Athanasius's theological concerns were thus embedded in an ongoing religious and philosophical shift toward transcendence and reduced access to the divine during the fourth century. His insistence on the security of revealed knowledge through the essential image rather than through creation was shaped by contemporary assumptions about divine transcendence, even if adapted dramatically to a Christian context by the affirmation of the incarnation of the transcendent one.

As a beleaguered, able bishop, Athanasius forged unity out of pluralism in tradition and Scripture, as well as in the church, by his devotional and epistemological focus on the incarnate Christ, the arresting solution to contemporary transcendence, dualism, and ecclesiastical disorder. Athanasius believed not in Eusebius's vision of divine power in salvation history, but in the fundamental act of God's incarnate Son to transform the fallen world and ensure salvation. By contrast with the cosmological optimism of Origen's pedagogical and individualistic soteriology, the strength of Athanasius's theology lay in its dramatic and corporate simplicity: the Incarnation was the clarifying and compelling key to alienation and disorder, chaos and death. This, however, was not

[29] Dörrie, 'Logos-Religion?'; P. Brown, *A Social Context to the Religious Crisis of the Third Century A.D.* (Berkeley, Calif. 1975); Williams, *Arius*, 199–214.

[30] On Manichaean epistemology and Christology, see E. Rose, *Die Manichaïsche Christologie* (Wiesbaden, 1979), 122. Mani proclaimed inherited wisdom to be inferior to a new revelation, in H. Ibscher (ed.), *Kephalaia* (Stuttgart, 1940), 8. For a comparison of Manichaean and Egyptian Christology, see my 'Arians and Manichees on Christ', *JTS* ns 40 (1989), 493–503.

[31] Questions regarding Arius's epistemology have been recently summarized and discussed in Williams, *Arius*, 106–7, 207f. For later Arian claims of knowledge, see Kopecek, 'Neo-Arian Religion', 172–4.

simply a clarification of biblical orthodoxy, but a powerful vision constructed to meet the spiritual and theological challenges of the fourth-century church.

ATHANASIUS AS THEOLOGIAN

Divine nature and will

Athanasius's defence of the essential unity of the Father and the Son represents a significant shift in Christian Platonic cosmology, usually analysed as combining a stricter application of *creatio ex nihilo*, the rejection of Greek hierarchical cosmology, and the embodiment of a soteriological rather than a cosmological view of the world. This has given rise to a special concern with the relationship between contemporary Platonism and Christianity in his thought.[32]

Throughout his works he uses the metaphysical commonplaces of divine transcendence from contemporary Platonism: '*ὁ ὑπερ πάσης οὐσίας, ὁ δημιουργὸς, ἀόρατος, ἀσώματος, ἄτρεπτος, ἁπλοῦς*.[33] Eternity, transcendence, and unity were fundamental definitions of divine being. As self-sufficient and ungenerated, God does not share any sort of essence with the created order; God is a simple, eternal essence to which nothing may be added.[34] Divine attributes are thus eternal. As often noted, Athanasius treats eternity not merely as a philosophical issue, but also as a religious one. His ontological presuppositions were translated directly into statements of faith: God's eternal being entails that revelation is constant and consistent; God does not change towards humanity:

> For the gods of the Greeks . . . are faithful neither in essence nor in promises . . . The God of all being one and truly God is faithful, saying,

[32] For a review of the literature, see Meijering, *Orthodoxy and Platonism*, 114f.; Ricken, 'Rezeption', 343–9; de Vogel, 'Platonism and Christianity', 61f.

[33] For *ὁ ὑπερ πάσης οὐσίας*, see *Gent.* 2. 6 (ed. Thomson, 6), 35. 3 (ed. Thomson, 94); for *ὁ δημιουργὸς*, see *Gent.* 2. 5 (ed. Thomson, 6); for *ἀόρατος*, see *Gent.* 29. 2 (ed. Thomson, 78), 35. 22 (ed. Thomson, 94); for *ἀσώματος*, see Gent. 29. 2 (ed. Thomson, 78); for *ἄτρεπτος*, see *Ar.* 1. 22 (ed. Bright, 23), 1. 35 (ed. Bright, 36), 1. 40 (ed. Bright, 41), 1. 28 (ed. Bright, 29), 2. 34 (ed. Bright, 103); for *ἁπλοῦς*, see *Ar.* 1. 28 (ed. Bright, 29).

[34] *Ar.* 1. 20 (ed. Bright, 21), 1. 28 (ed. Bright, 29); *Decr.* 11 (*PG* 25. 441–4).

'See me, see that I am' and 'I change not'. (*Ar.* 2. 10 (ed. Bright, 78))

God's faithfulness is due to the ontological properties of divine nature.[35]

The perfection of God's essence includes a sense of priority and causality, but in the entirely different and mysterious quality of his being. Origen and Eusebius used Exodus 3: 14 to ground a description of God as the source of all existence; likewise the Arians used 'ὁ ὤν' to identify the Father as being itself, which meant that the Son must originate from nothing.[36] Athanasius defined 'ὁ ὤν' as the eternal, hence transcendent, perfection of the divine essence, but not as absolute source, lest it exclude the Son; he even rhetorically and untraditionally claimed that the Son may be defined by 'ὁ ὤν'.[37] Arguing that divine essence was by nature eternally generative as a fountain, he maintained that the Father was by definition the source, but only within the divine essence.[38] He thus claimed that anything outside the eternal unity of the Godhead must be accidental; otherwise it would compound the simple divine essence. The Father communicated the whole of divine essence to the Son.[39] The eternal self-expression of God was therefore Father, Son, and Holy Spirit. Yet, this was not a hierarchy of decreasing essences intervening between the Father and the world, for the only division lay between uncreated and created being.

Like his predecessors, Athanasius used the terms *φύσις* and *οὐσία* to describe the divine essence. As recently discussed by Stead, in spite of the pressures of the Arian controversy, he gave them no technical meaning apart from the sense of essential communication between Father and Son.[40] According to Athanasius, divine essence could not be conceived of in any material way or by material analogy, for it was incomprehensible except by the titles 'God' or 'Father', which name rather than describe it.[41] He gave content to divine nature by reference to qualities proper to eternal being. He thus shared the contemporary Christian

[35] Cf. discussion in de Vogel, 'Platonism and Christianity', 61 f.

[36] *Ar.* 1. 24 (ed. Bright, 25); *Ur.* 7.4 (ed. Opitz, 14–15).

[37] *Ar.* 1. 12 (ed. Bright, 14); *Decr.* 22 (*PG* 25. 456).

[38] *Ar.* 1. 19–21 (ed. Bright, 20–2); cf. *Ep. Serap.* 1. 28 (*PG* 26. 609).

[39] E. P. Meijering, 'Athanasius on the Father as the Origin of the Son', in idem, *God Being History*, 96; Stead, *Divine Substance*, 263.

[40] *Ar.* 2. 41 (ed. Bright, 110), 2. 45 (ed. Bright, 114); Stead, *Divine Substance*, 233.

[41] *Decr.* 22 (*PG* 25. 453–6); *Syn.* 34 (*PG* 26. 753).

modification of Middle Platonic participation language. Whenever he used the term μετέχω, he modified it to express the essential unity of the Father and the Son, or used it alone strictly of created beings. In the fourth century, Christian participation implied a relationship of grace. Hence the Son, as 'begotten', had a 'whole participation'; God was not participated in except wholly, by the Son, for God does not share his essence by affection or division.[42] The argument thus functioned to preserve the simplicity of divine nature and the divinity of the Son. Athanasius used the term 'essence' (οὐσία) within the Godhead to describe the unique identity of divinity, rather than shared qualities which may alter or will which may change.[43] Agreeing with critics that God is one and may not have another or foreign essence proper to himself, he argued that the Son and the Father as divine must have one essence.[44]

From these Christian Platonic assumptions of divine simplicity and eternity, Athanasius interpreted scriptural language about God. For example, against the later Arian use of ἀγένητος to define divine essence, he argued that 'Father' was the essential definition of God; this title was simple and scriptural, and anchored in the testimony of the Son.[45] 'Fatherhood' therefore described God's essential self-expression in the Son, and revealed an interior relationship: God was always Father, but not always Creator.[46] This point, as in Irenaeus and the *Gospel of Truth*, was of course directly linked to the eternity and soteriological revelation of the Son: Christians know the 'Father' intimately through revelation by the 'Son', not like Greeks by analogy through creation.[47] Yet scriptural, hence divine fatherhood must be defined properly by the qualities of eternity and essential communication, not temporal priority. 'Father' also described the generative nature of divine being, which expressed itself eternally as does light by its radiance. Within the relationship of the Father and the Son, he therefore emphasized the dynamic process of essential

[42] *Ar.* 1. 15–16 (ed. Bright, 16–18), 1. 28 (ed. Bright, 29); *Syn.* 48 (*PG* 26. 777–80), 51 (*PG* 26. 784). See also Meijering, *Orthodoxy and Platonism*, 62f. Williams notes Arius's rejection of participation in *Arius*, 222–3.

[43] *Syn.* 34 (*PG* 26. 753); *Ar.* 1. 20 (ed. Bright, 21); *Afros* 8 (*PG* 26. 1044).

[44] *Syn.* 50 (*PG* 26. 781–4).

[45] *Ar.* 1. 28 (ed. Bright, 29), 1. 33–4 (ed. Bright, 34–7); *Decr.* 22 (*PG* 25. 456).

[46] *Ar.* 1. 29 (ed. Bright, 30), 1. 21–2 (ed. Bright, 22–3).

[47] *Ar.* 1. 33–4 (ed. Bright, 33–5); cf. Irenaeus, *Haer.* 4. 6. 6.

communication proper to shared, divine nature. Whatever is said of the Father may also be said of the Son, except that he may not be given the title 'Father'.[48] Such a description excluded aspects of priority or volition which other theologians wished to apply to a notion of divine fatherhood, but ensured a continuity of divine essence flowing from Father to Son like a fountain or spring.[49] Unlike Eusebius or Arius, who modified such language by reference to will, Athanasius defended eternal, essential generation as proper to God. Athanasius's defence and exegesis of the primary scriptural title of God as 'Father', however, rested firmly on the contemporary metaphysical definitions of eternal generative nature.

Athanasius also defined divine goodness by reference to contemporary categories. Like his predecessors, he used the Platonic description from the *Timaeus* as a matter of course: 'ὁ δὲ τῶν ὅλων θεὸς ἀγαθὸς καὶ ὑπερκαλος τὴν φύσιν ἐστί· δìο καὶ φιλὰνθρωπος ἐστιν· ἀγαθῷ γὰρ περὶ οὐδενὸς ἂν γένοιτο φθόνος.'[50] This creative goodness, however, was reflected imperfectly in the world, which was created to reveal the Father, the source (πήγη) of goodness.[51] Unlike Clement and Eusebius, he did not qualify these formulas to underline God's freedom or will in goodness or creation. In reaction to Arian teachings on divine will and in line with some contemporary Platonists, he argued that God was above will and freedom in relation to goodness:

> To say of the Son he might not have been is an impious assumption reaching to the essence of the Father as what is his own might not have been. For it is the same as saying, the Father might not have been good. . . . For as in the case of radiance and light one might say, there is no will preceding radiance, but it is its natural offspring. (*Ar*. 3. 66 (ed. Bright, 219))

This passage's reference to radiance and goodness is in direct contradiction to Eusebius's description in *DE* 4. 1. As recently discussed by Stead, Athanasius here reflects a general discomfort

[48] *Ar*. 3. 4 (ed. Bright, 157–9); *Syn*. 49 *PG* 26. 781).

[49] *Ar*. 2. 41 (ed. Bright, 110–11), 3. 66 (ed. Bright, 219). Cf. Stead, *Divine Substance*, 263–4.

[50] *Gent*. 41. 13f. (ed. Thomson, 112–14); *Ar*. 2. 77 (ed. Bright, 147). Cf. Plato, *Timaeus*, 29E.

[51] *Gent*. 35. 12–14 (ed. Thomson, 94), 7. 3 (ed. Thomson, 16); 40. 15 (ed. Thomson, 110), 41. 5–8 (ed. Thomson, 112–14). Cf. Eusebius, *DE* 4. 1 (*GCS* 23. 150–1).

with the concept of will, since it is changeable and hence inappropriate to his definitions of divine nature.[52] If divine nature had any kind of arbitrary will, it would no longer be eternal, but equivalent to the changeable and material sense of *ῥοπή*. Human beings have free will, but animals are below choice, and divinity is above it.[53] Divine will, therefore, is eternal, like divine nature; hence God acts and loves the good by nature.[54] Although, as Stead points out, philosophically speaking, Athanasius may pose an unreal antithesis between nature and will, the proper question concerning eternity or temporality, it is important to note that theologically this definition is fundamental to his understanding of divinity and soteriology, as well as to his arguments against his opponents. Will in general (*βουλή*) for Athanasius is episodic and external. Although with regard to the generation of the Son, he defined divine will as eternal and reflective of divine essence, in reference to divine activity in the world it is a temporary act.[55] Thus, as in Origen, decision and act are one process in divine nature, so no duality exists. By contrast with Eusebius and Porphyry, the emphasis is on inner freedom, as opposed to outward power; as in Origen and Plotinus, *θέλημα* defines the interior freedom of God.[56]

Equally important, in these passages the Son is described as the eternal desire or good pleasure of God: '*οὐκοῦν μὴ λεγέσθω θελήματος δημιούργημα ὁ υἱὸς . . . ἀλλὰ βουλὴ Ζῶσα, καὶ ἀληθῶς φύσει γέννημα, ὡς τοῦ φωτὸς τὸ ἀπαύγασμα.*'[57] This description of divine will is critical to his defence of the essential and eternal generation of the Son. The Son is a product of the love and desire of God, and hence eternal; a similar statement was made by Origen. Yet, because of his suspicion of divine will and the controversy over the generation of the Son, in Athanasius the Son became literally the will of God. God did not have two wills, one preceding the Son and bringing him into existence, one being the Son.[58] Here his argument is directed against an Arian sense of creation of the Son; hence divine will is given

[52] Stead, 'Freedom of Will', 257. [53] *Ar.* 3. 18 (ed. Bright, 173).

[54] *Ar.* 1. 52 (ed. Bright, 54), 1. 20 (ed. Bright, 21–2); Williams, *Arius*, 228.

[55] *Ar.* 2. 24 (ed. Bright, 93), 3. 10 (ed. Bright, 163–4); *Inc.* 44 (*SC* 199. 424f.); cf. Stead, 'Freedom of Will', 256–7).

[56] Compare *Ar.* 3. 66 with Plotinus, *Ennead* 6. 8. 13; cf. Meijering, *Orthodoxy and Platonism*, 80.

[57] *Ar.* 3. 67 (ed. Bright, 220).

an eternal character identified with the Son in the same way that the Son is eternally divine Wisdom or Word. At times he retains a separate sense of paternal will in creation and redemption, reflecting traditional language of the agency of the Word as much as a hierarchy of divine action.[59] More commonly, however, the Son and the Father, being of one nature, have one will, with little sense of separate agency: the Son represents the Father and is the paternal will.[60]

These explanations of divine will were, of course, embedded in controversy concerning the generation of the Son; but they do reveal Athanasius's assumptions concerning divine will and nature. The latter are further clarified by looking at his doctrine of creation. Unlike his predecessors or his contemporaries Eusebius and Methodius, he did not appeal to the sheer power of God to justify the shape of creation or redemption. The doctrine of *creatio ex nihilo* had developed against a Gnostic or Manichaean sense of an evil or fatalistic creation by asserting the power and will of God. Irenaeus, and more recently and sharply, Methodius, and Eusebius all underlined power and goodness in divine activity. Athanasius, however, focused on the contrast between uncreated eternal and created contingent natures. Thus divine nature and will are good and eternal, yet the effect of divine will in creation is surprisingly weak. In Athanasius's early works, God is said to create by will; but his overall account is more a matter of God creating 'out of a limbo of unreal being'.[61] God, by will in creation, did not thereby determine the structure of reality for salvation—for example, the pedagogical development of free beings as in Irenaeus—but organized the circumstances to compensate for the fragility of created being. Depending on divine will as foreknowledge rather than on personal response, he noted that God as good not only created the world, but, like an architect, willed the repairs for the house before he built it.[62] In *On the Incarnation* God as Creator appears not only metaphorically but actually to be a builder with faulty material and as an exasperated king.[63]

[58] *Ar.* 3. 61 (ed. Bright, 214), 2. 31 (ed. Bright, 101).

[59] *Ar.* 2. 7 (ed. Bright, 75), 3. 31 (ed. Bright, 185), 3. 60 (ed. Bright, 213), 2. 64 (ed. Bright, 134).

[60] *Ar.* 2. 24 (ed. Bright, 93); cf. *Ar.* 3. 10 (ed. Bright, 163–4).

[61] Stead, *Divine Substance*, 237.

Obviously, this is not the first time that divine will has been limited by contingent nature in Christian cosmology; but Athanasius did not appeal to a soteriological sense of divine power or the necessary interaction between God's plan and human freedom, but rather to the eternal foresight of divine nature. The pull of non-being and dissolution may be stronger in Athanasius because his ontology is discontinuous, and the generic instability of created being rendered it liable to return to nothing. Yet, curiously, at the same time that divine will became primary in creation, it lost power.

In Irenaeus, Origen, and Eusebius, creatures, while contingent, continued to exist within a world shaped by God's purpose. Whatever their condition, *αὐτεξούσιος* implied a moral freedom and essential likeness to God; hence creatures received power and salvation by nature and will. For Athanasius, products of divine will were subject to motion by definition, as *τρέπτος*. God desired their immortality in a secondary sense: '*ὁ θεὸς τὸν ἄνθρωπον πεποίηκε, καὶ μένειν ἠθέλησεν ἐν ἀφθαρσίᾳ*.'[64] Although Athanasius insisted that creation was good, the created world, by contrast with the transcendent order, had an independent quality of temporality and contingency. Like Origen, Athanasius identified evil as non-being or illusion—that is, the turning away from the reality and true being of God.[65] Yet, the stark contrast between the eternal transcendent nature and the contingent creation increased the fragility of the relationship between God and humanity, making the created being by motion liable to return to nothing. Thus, to protect the divine image of rationality against its essential mutability, God granted creatures the Law and a place, Paradise; yet they inevitably turned away, and continued to fall into corruption.[66] Creation, in spite of divine will, was subject not only to moral change or disorder, but also to an essential instability. God therefore became subject to his own justice in the Fall; to restore fallen humanity and preserve the integrity of the design, it was necessary for him to overcome his own Law.[67]

Like those of his predecessors, Athanasius's cosmological dis-

[62] *Ar.* 2. 77 (ed. Bright, 147). [63] *Inc.* 13. 14f. (*SC* 199. 310–12).

[64] *Inc.* 4. 4 (*SC* 199. 276).

[65] *Gent.* 2. 1–3 (ed. Thomson, 4–6), 4. 18–20 (ed. Thomson, 10). Cf. Gregg and Groh, *Early Arianism*, 178.

[66] *Inc.* 3. 24–40 (*SC* 199. 270–4), 5 (*SC* 199. 278f.).

cussions reflect religious and theological concerns; but the latter have subtly shifted. The innate weakness and instability of the creation, by stark contrast with the stability and eternity of transcendent nature, overshadow salvation history, and serve to highlight the need for the Incarnation. Just as will was unable to create a perfect creature, so divine will alone would not be able to redeem it; will was an exterior power which could not stabilize or transform fallen human nature, hence the communication of divine nature was needed.[68] This untraditional assumption seems directly related to the changing cosmologies of the fourth century: for Athanasius, as for Alexander and Methodius, the human tendency to corruption could only be stayed by incarnation.[69] The primary theological issue was no longer fatalism, but transcendence and the alienation of material existence. Creation remained in bondage in part because it was external to God as created, and therefore needed to be transformed internally and essentially.[70] The communication of divine nature therefore outweighed the emphasis in older Christian models on moral freedom or the sovereignty of divine will. These elements remain, but in muted form.

As argued by Roldanus and by Gregg and Groh, these reflections on divine will were shaped not only by a defence of incarnation, but also by the emerging asceticism of the fourth century.[71] Although Athanasius retained some of the traditional senses of *πρόνοια*, he also shared the sharpened ascetic sense of intervention and deliverance by God away from society, rather than the sense of a broad leaven of divine progress within history. In religious contexts he continued to refer to the will of God as a backdrop for human efforts. God worked in varied and lovingly kind ways to redeem individuals:

For God is always the same; but people have come into existence through the Word, when the Father himself willed it, for God is invisible

[67] *Inc.* 6–7 (*SC* 199. 282–8). See Mühlenberg, 'Vérité et bonté'.

[68] *Inc.* 44 (*SC* 199. 424–5): God cannot restore natures by *fiat*. Cf. *Ar.* 2. 67–8 (ed. Bright, 137–9).

[69] Alexander, 'On the Soul and Body and Passion of the Lord', *ANF* 14, 356–63. On this shift of emphasis in the late third century, see also Brown, *Body and Society*, 90–202.

[70] *Gent.* 46. 47–51 (ed. Thomson, 128–30); *Ar.* 2. 76 (ed. Bright, 147).

[71] Abba Poimen said, 'It is our wills that play the demon with us', quoted in Brown, *Body and Society*, 226; cf. Gregg and Groh, *Early Arianism*, 177 f., Roldanus, *Le Christ*, 371.

and inaccessible to created things, and especially to people on earth. When the weak call upon him, when pursued they beg for rescue, when under injustice they pray, the invisible being a lover of humanity shines forth upon them with his good works which he does through his Word. (*Ar.* I. 63 (ed. Bright, 66))

As in this passage, there was usually little reference to the mercy or guidance of the Father apart from Christ.[72] The problems of the single-hearted ascetic perhaps pushed his thought away from salvation history toward a stronger, more decisive sense of divinization through the Word from the transcendent Father. The defence of providence was no longer important, since history was simply irrelevant to those struggling to live above nature and like angels in and through the power of God.[73] The primary sense of divine character was of transcendent stability.

Athanasius's reflections on divine nature and will were thus anchored in contemporary concerns regarding creation, transcendence, and ascetic practice. Together, these concerns pushed his largely traditional theology towards a different set of foci from what are found in some of his predecessors. The separation of nature and will in Athanasius's doctrine of God enabled him to distinguish between the essential generation of the Son and the creation of the cosmos; this has often been praised as a critical step in the liberation of God from mere creator to being the Father of Jesus Christ.[74] This was an important aspect of his thought; but it was clearly linked not only with soteriology and religious concerns but also with cosmological assumptions about nature and will in God. In making this theological move Athanasius did not evade Platonic categories any more than did his predecessors. Indeed, he altered the earlier central Christian doctrines of creation and direct connection of God to the world to underscore eternity rather than power. To describe God's faithfulness as grounded in eternal nature rather than discerned in history is not a more 'Christian' step, but a continuing synthesis of faith and culture.[75]

[72] Cf. *Vita Antonii*, 48 (*NF* 4. 209); Letter 10. 4 (*NF* 4. 528).

[73] *Inc.* 19 (*SC* 199. 334). However, K. Torjesen defends the importance of the pedagogical role in 'The Teaching Function of the Logos', in Gregg (ed.), *Arianism*, 213.

[74] Ricken, 'Nikaia', 340.

[75] See de Vogel's criticism of Ricken in 'Platonism and Christianity', 4.

Human nature and will

Athanasius's reflections on anthropology reveal a similar mix of ascetic interests, polemical refutations, and soteriological issues. Roldanus described his polemical and pastoral interests as centred in an incarnational Christology shaped by asceticism; Kannengiesser detected his ascetical interests in the simplification and transformation of Origen's anthropology.[76] These concerns led him to underline the instability of human nature, rather than its self-determination, and to define humanity as a created *νοῦς* which, through the Incarnation, received revelation and incorruptibility, thereby enabling it to be made godlike.[77] These theological points supported the centrality of incarnation and revelation, but led to a greater emphasis on the weakness of the human will, which distinguished his position from that of his predecessors.

In general his anthropology rested on the exploration of *creatio ex nihilo* within the assumptions of Platonic ontology. Thus, creation was utterly contingent and separate from divine nature, and, as created, was subject to corruption and change in contrast to the eternity and unchangeability of God. Some ambivalence existed as to the status of humans as created images. In *Against the Gentiles* he gives an optimistic Platonic anthropology: the *νοῦς* was rational, immortal, and essentially capable of the vision of God, even if fallen. However, the created *νοῦς* was by nature mobile (*εὐκίνητος*), and so could be distracted from true reality and tarnished by bodily passions.[78] This problem of changeability was addressed in the more directly soteriological *On the Incarnation.* Here the creation of humans from nothing was one step, and the bestowing of the image of God—that is, rationality—another:

> He gave them an additional grace . . . making them in his own image and giving them a share in the power of his own Word . . . so that being made rational, they might be able to remain in blessedness, living the true life of the saints in paradise. Also, knowing that human will could

[76] Roldanus, *Le Christ*, 6; C. Kannengiesser, 'λόγος et νοῦς chez Athanase d'Alexandrie', *SP* 11 (1967), 199–202.

[77] A. Louth, 'The Concept of the Soul in *Contra Gentes—De Incarnatione*', *SP* 13 (1975), 227–31; Meijering, *Contra Gentes*, 24–7.

[78] *Gent.* 4. 8 (ed. Thomson, 10).

turn either way, in anticipation he secured the grace they had been given by means of a law and a place. (*Inc.* 3 (*SC* 199. 270))

Although he kept the traditional definition of the mind as the image of God, he heightened the contrast between the mind, as created, and the uncreated God: their connection was a grace.[79]

The instability of created nature, which underlay the traditional language of participation and divinization, created a number of new problems in Alexandrian anthropology. First, Athanasius simplified the anthropology of Origen by defining humanity as an embodied *νοῦς*, rejecting the myth of pre-existence with the transformation of *νοῦς* to *ψυχή*.[80] Humans remain themselves, created, embodied *νοῦς*, even if estranged from God by disobedience and death. This followed the emphasis of Methodius, for example, on the material origin and purpose of humans in a critical response to Origen's doctrine of pre-existence.[81] Equally important, it strengthened the centrality of the Incarnation as the revelation of God as embodied mind to the embodied mind; the noetic act becomes the act of faith: 'For our salvation he had compassion to the extent of being born and revealed in a body.'[82] The encounter with the incarnate Word thus led to re-creation rather than instruction, since divine knowledge linked the mind to God, staying sin and bodily distraction. Curiously, Platonic contemplation remained central in Athanasius's soteriology, albeit transformed by its use in an incarnate setting: by Christ's dwelling in the flesh, sin might be perfectly expelled from the flesh, and we might have a free mind.[83]

Clearly, his account of the alienation of humanity and salvation centred on the tension between mind and body. Although he affirmed the single material creation of humanity, in line with contemporary ontological assumptions, he located error in the tension between rationality and the material: in the Fall a mental lapse through regard for the body estranged humanity from

[79] *Inc.* 3 (*SC* 199. 274), 4 (*SC* 199. 276–7). Kannengiesser, 'Athanasius of Alexandria', 109. For a contrast between Origen and Athanasius on *νοῦς*, see Roldanus, *Le Christ*, 370.

[80] Kannengiesser, '*λόγος*', 200f.

[81] C. Kannengiesser (ed.), *Sur l'Incarnation du Verbe*, *SC* 199. (Paris, 1973), 75: Methodius, *Res.* 1. 60, 6; *Sym.* 1. 4.

[82] *Inc.* 4. 12–17 (*SC* 199. 276); see phrase of Kannengiesser, *Sur l'Incarnation*, *SC* 199. 156.

[83] *Ar.* 2. 56 (ed. Bright, 126).

God.[84] Thus, rationality was tarnished by inevitable distraction, which lay in the natural instability of created being rather than in deliberate disobedience. Adam may be contrasted with the later saved Christian, therefore, as possessing only an external grace, which could not overcome his created instability; only incarnation, not instruction, can destroy the passions of the body thereby allowing one to become impassible and free:

For the Word being clothed in the flesh . . . every bite of the serpent began to be utterly staunched by it, and whatever evil came from the motions of the flesh, to be cut away. . . . And these being destroyed from the flesh, we all were liberated through the kinship of the flesh. (*Ar.* 2. 69 (ed. Bright, 139))

In this passage he refers to the instability after the Fall; but actually, only the degree of instability increased, an innate physical weakness remaining the cause of estrangement.[85] The essential instability of the created body thus outweighed the disobedience of the will, in contrast to earlier descriptions of the Fall.

As noted by Roldanus and by Gregg and Groh, this focus probably reflected Egyptian and ascetic concerns regarding the corruptibility of the body.[86] Ascetics focused on the weak body as the very place to practice and achieve an imitation of divine stability. Brown speculated recently that Christological issues regarding incarnation were particularly significant to those who sought to achieve physical sanctity as part of moral ἄσκησις.[87] Within the chaotic society of the fourth century, the unblemished, disciplined body of the ascetic became the direct, physical sign of the power of God. However, in Athanasius's perspective, the transformed body was also collective, and thus a defining characteristic of the church. The reality of physical restoration was therefore essential to the unique, sacred identity of the church within Late Roman society.[88]

Because of his heightened theological focus on contemplation and physical vulnerability, Athanasius's reflections on human action were necessarily different from those of earlier authors.

[84] *Gent.* 2. 21–6 (ed. Thomson, 6), 3. 5 (ed. Thomson, 8).

[85] *Ar.* 1. 51 (ed. Bright, 53), 2. 56 (ed. Bright, 126), 2. 68–9 (ed. Bright, 138–40). Cf. Roldanus, *Le Christ*, 117f.

[86] Roldanus, *Le Christ*, 352; Gregg and Groh, *Early Arianism*, 131.

[87] Brown, *Body and Society*, 236.

[88] *Ar.* 1. 42–4 (ed. Bright, 42–6).

Within his noetic anthropology, will was not an important faculty, but was rather a largely negative expression of changeable nature.[89] Given the emphasis on contemplation and corruptibility, *προαίρεσις* became a means of virtue rather than a significant aspect of the image of God which ensured human progress toward divinization as in Irenaeus, Origen, or Eusebius. The issue for Athanasius was corruptibility and alienation, rather than fatalism. Resolving issues of Christian Platonism, freedom therefore consisted in deliverance from bodily corruption or mental distraction which hindered the free will; this could not be achieved through external teaching, but only through a transformation of nature.[90]

Yet, if his more formal theological reflections emphasized the necessity of the Incarnation, his practical and spiritual advice to others consisted of a traditional mixture of moral exhortation and appeals to Christ as exemplar. He defended the essential moral freedom of the human will thus:

For the world is like a sea to us . . . We float on this sea as with the wind through our free will, for everyone directs his course according to his will and either under the pilotage of the Word enters into rest, or laid hold only on pleasure, he suffers shipwreck. (Easter letter 347 (tr. Robertson, 547))

Salvation and Christian life for Athanasius were not therefore wholly passive matters: because the will may incline either way, one must take pains to seek virtue, gain grace, and persevere.[91] This endurance required of all people in all states, on all levels from virgins to married people, is modelled on the example of Christ and is strengthened by the sacraments:

For in sowing he did not compel the will beyond the power. Nor is mercy confined to the Perfect, but it is sent down among those who occupy the middle and third ranks so that he might rescue all to salvation. To this intent he has prepared many mansions with the Father, so that although the dwelling place is various in proportion to the advance

[89] Stead, 'Freedom of Will', 250; Kannengiesser (ed.), *Sur l'Incarnation*, *SC* 199. 152; see above on Plotinus, p. 30.

[90] *Ar.* 2. 56 (ed. Bright, 126), 3. 33 (ed. Bright, 187).

[91] Festal letter (329) .5 (*NF* 4. 508); letter to Dracontius 7–8 (*NF* 4. 560) on the necessity to advance; *Ar.* 3. 25 (ed. Bright, 180).

in moral attainment, yet all of us are within the wall and all of us enter within the same fence. (Easter letter 338 (tr. Robertson, 592))[92]

Significantly, these traditional statements about the moral life remained a minor spiritual theme in Athanasius's pastoral writings, rather than a theological focus, as in Irenaeus or Origen.

Soteriology in Athanasius thus centred on the weaknesses of physical life and deliverance therefrom through divinization. In a real sense, divinization begins now, in the transformed life of the Christian: those who are virtuous in Christ become 'dead to the world' and 'detached from the sensible'.[93] The goal of the saint was thus a firmly ascetic one: in cleansed flesh, virtue increases so that one no longer lives on earth and has no temptation, particularly of the sexual variety.[94] Divinization (*θεοποίησις*) therefore involves the whole person, body and mind, rather than escape from the body, but in line with the limits of human created nature; for one cannot become like God who is truly unchangeable in essence, but only by progress in virtue imitate God.[95] Participation is therefore equivalent to grace in the sense of an external, transforming relationship with God which allows a certain sharing in the power of God, but not of an ultimate transformation into divinity. Yet the language of Athanasius remains shaped by the condition of living away from a historical, material setting and moving toward the noetic, eternal world. Physical redemption is the key to the solution of the most troublesome aspect of spiritual liberation for a Christian Platonist.

The old language of moral training was therefore set firmly in a new ascetic framework with a focus on the physical body. It is difficult to decide whether this is a justification of the Incarnation or establishes its necessity; and perhaps it is artificial to even try to decide. In his emphasis on mental and physical transformation, Athanasius seems to be addressing lingering problems concerning created human instability in the older Christian Platonic model; a transformed body would steady the disobedient will.

92 Festal letter (338) .4 (*NF* 4. 529); Festal letter (330) .5 (*NF* 4. 511); Festal letter (332) .3 (*NF* 4. 516); Festal letter (333) .4 (*NF* 4. 518). Stead has commented that these passages reflect the persisting importance of Christ as moral exemplar in 'Freedom of Will', 255.

93 *Ar.* 3. 52 (ed. Bright, 205).

94 *Ar.* 2. 69 (ed. Bright, 140); cf. *Ar.* 3. 34 (ed. Bright, 188–9), 2. 56 (ed. Bright, 126).

95 *Ar.* 2. 70 (ed. Bright, 141).

Yet his emphasis on living like angels or apart from the world through the transformation of the physical seems hardly less dualistic than that of his predecessors.[96] Indeed, as *τρέπτος* rather than *αὐτεξούσιος*, human nature is alienated more profoundly from history or the world, as well as from eternal divine nature, than it is in Irenaeus or, ironically, even Origen. The body, if the locus of transformation, has become to an even greater extent the principal distraction. The liberation of human nature through the divine incarnation may be against the logic of contemporary categories, but its power rests squarely on overcoming contemporary concerns. The stabilization of body and mind solved problems critical to the Manichees and Platonists alike: those of physical instability and mental distraction. Athanasius affirmed the created, historical world as the place for divine and human activity; but the power of salvation lay not in the progress of history, but in the incarnational event which created islands of redeemed humanity in the fallen world.

Christology

Athanasius's defence of the essential divinity of the incarnate Son has often been described as 'kerygmatic' or 'soteriological' rather than cosmological: the Word as fully God incarnate becomes the true Saviour, not the mediator. If incarnation is the key to the theological relationship between God and human nature, nevertheless, its shape and function remain profoundly linked to the great contrast between transcendent divinity and created humanity in the fourth century. Rather than seeing this soteriology as shaping all aspects of his thought, we need to explore the connections between transcendent deity and ascetic humanity within his theological model. If in regard to Origen and Eusebius I outlined the Christological impact on contemporary cosmology, here a balance is needed to outline the cosmological impact on Athanasius's Christology, which has been variously described as anything from 'essentialist' to 'biblical'.[97] His model answered some traditional questions in a compelling way, yet raised new

[96] See Kannengiesser, 'Athanasius of Alexandria', 110, for where he claims that Athanasius has a less dualistic anthropology; '*λόγος*', 154–6 for the glorification of the somatic state.

[97] Young, *Nicaea to Chalcedon*, 69, 83; Ricken, 'Nikaia', 348f.; Kannengiesser, 'Athanasius of Alexandria', 112f.

problems, including an ambiguity concerning the reality of the humanity of Christ, sometimes focused on the question of the presence of a human soul in Christ.[98]

In the context of fourth-century theological controversies and in line with his definition of the absolute simplicity of divine nature, Athanasius argued that the divine nature of the Son must be exactly like that of the Father:

ὁ δέ γε υἱὸς, οὐκ ὢν ποίημα, ἀλλ' ἴδιος τῆς οὐσίας τοῦ πατρὸς, ἀεὶ ἔστιν· ἀεὶ γὰρ ὄντος τοῦ πατρὸς, ἀεὶ εἶναι δεῖ καὶ τὸ ἴδιον τῆς οὐσίας αὐτοῦ, ὅπερ ἐστὶν ὁ λόγος αὐτοῦ καὶ ἡ σοφία. (*Ar.* I. 29 (ed. Bright, 31))

As can be seen in this passage, whatever the title of the Son, his nature had to be described in a way appropriate to the unchanging, transcendent divine essence. Consistent with his theological method, Athanasius accepted the titles of 'Father' and 'Son' as axioms of faith. These were not subject to human interpretation, since nature determined the meaning of language; rather, they revealed God's essential, eternal being. Thus, the divine, unchanging Father required an eternal, co-essential Son, to be consistent with the definitions of divine essence. If the Son were not eternal, a change would have occurred in the Godhead, and God would not be eternally Father or presumably true God.[99] Some later theologians with suspicions of Alexander's modalism akin to those of Arius, challenged this exegesis on the grounds that one does become a father at some point, asking whether they were co-existent or, facetiously perhaps, 'brothers'?[100] Athanasius rejected such questions as unworthy of the eternal simplicity of divine essence and the integrity of revelation; if the Son were not of the unchanging divine essence of the Father, his relationship would be an exterior one of adoption or mere participation, like that of the rest of creation.[101] Since he rejected participation as inadequate, from his earliest work (*Against the Pagans*), the essential identity of the Father and Son would seem

[98] M. Richard, 'Saint Athanase et la psychologie du Christ selon les Ariens', *Mélanges de science religieuse*, 4 (1947), 5–54; Stead, 'Scriptures and Soul of Christ'; A. Louth, 'Athanasius' Understanding of the Humanity of Christ', *SP* 16.2 (1985), 309–18.

[99] *Ar.* I. 15–16 (ed. Bright, 16–17), I. 35 (ed. Bright, 36–7), I. 39 (ed. Bright, 40–1).

[100] *Ar.* I. 14 (ed. Bright, 15), I. 22 (ed. Bright, 24).

[101] *Ar.* I. 15 (ed. Bright, 16–17), 3. 17 (ed. Bright, 171f.), 3. 19 (ed. Bright, 173).

to be an axiom of his thought in line with the Wisdom Christology of some Alexandrians.[102] To be the Son, he had to be the natural, hence eternal, offspring of the Father.[103]

Following Origen and Alexander, Athanasius argued that the Son as the Word and Wisdom of God was eternally begotten. Perhaps by analogy with eternal creation or the Manichaean assertion of co-existent principles, this idea was criticized in the early fourth century by Arius as compromising divine simplicity; as discussed above, Eusebius emphasized the will of God in generation in order to defend divine freedom. Athanasius, however, argued that such a process reflected the perfection of divine nature: because the Father was ἀπαθής and ἁπλοῦς, the Son had an eternal, impassible generation. Like Alexander, Athanasius drew on the metaphors of Scripture, together with the qualities of divine nature. The Son was eternally begotten as radiance followed light; because the Father was always wise and rational, he always had his Son.[104] To an extent, eternal generation in Athanasius's arguments was circular so as to safeguard not only the simplicity of the Father, but also the divinity of the Son. In order to be divine, the Son must be eternal; in order for God to be 'Father', the process of generation must be eternal and essential. He thus interpreted 'begotten' in line with the qualities of divine nature, rather than by human analogy; thus 'begotten' was above decision, and meant a whole participation—that is, essential communication.[105]

The Son's nature was therefore not determined by his priority or proximity to the Father, but by their shared divine essence. Athanasius ridiculed the idea of the Son receiving divine qualities by appointment or will: will implied an external relation appropriate to contingent creatures, and this could not convey the divine—that is, eternal and unchanging—qualities.[106] Scriptural titles such as Word, Wisdom, or Image also supported his eternal, essential divinity. As the image of God, the Son had to be eternal and part of God's nature; if he were secondary or change-

[102] *Gent.* 46 (ed. Thomson, 130). Cf. Meijering, *Contra Gentes*, 149.
[103] *Ar.* 1. 19–20 (ed. Bright, 21), 1. 29 (ed. Bright, 30–1), 2. 30 (ed. Bright, 99).
[104] *Ar.* 1. 17 (ed. Bright, 18–19), 3. 28 (ed. Bright, 183), 1. 25 (ed. Bright, 26).
[105] *Ar.* 1. 21 (ed. Bright, 22–5); *Syn.* 51 (*PG* 26. 784).
[106] *Decr.* 13 (*PG* 25. 445–8); *Syn.* 36 (*PG* 26. 756–7); *Exp. Fid.* 3 (*PG*. 25. 197).

able, he could not be the true image of the Father.[107] Like Alexander, Athanasius rejected any sense of inferiority in the Son as image, for he was the epistemological tie between God and creation.[108] Indeed, as image, all qualities of divine nature must be attributed to him. Athanasius thus followed traditional Wisdom Christology in describing the essential likeness of the Father and the Son even in a work not explicitly against the Arians:

> Being the good offspring of the Good and true Son, he is the power of the Father and his Wisdom and Word; not by participation nor are these properties added to him from outside in the way of those who participate in him . . . but he is absolute wisdom [αὐτοσοφία], absolute word [αὐτολόγος], the absolute power [αὐτοδύναμις] of the Father. (*Gent*. 46 (ed. Thomson, 46))

Unlike his contemporary, Eusebius, Athanasius in this passage did not emphasize the priority of the Father in the αὐτο-categories, but rather the Son's essential right to them as mediator and image of the Father.

However, even this traditional language of mediation often disappeared in the polemical definitions which he argued against the Arians. All but the title 'Father' could be used of the Son, for whatever was said of the Father could be said of the Son from a creature's perspective.[109] Whatever categories were applied to the Father should therefore be used to understand the nature of the Son: if God was not caused by will, how foolish to consider that the Son was![110] By identification of the Son's nature with the divinity of the Father, older distinctions were modified, if not rejected.

Yet, as Stead has shown, Athanasius retained a certain traditional hierarchy between the Father and the Son. He used ὁμοούσιος cautiously, and only in the context of the Nicene creed.[111] Like Origen, he continued to use οὐσία and φύσις interchangeably to express a common essence.[112] On the one hand, he

[107] *Ar*. 1. 22 (ed. Bright, 23), 3. 10–11 (ed. Bright, 163–5).
[108] *Ar*. 2. 78 (ed. Bright, 148). [109] *Ar*. 3. 4 (ed. Bright, 157).
[110] *Ar*. 3. 63 (ed. Bright, 216); cf. *Ar*. 2. 35 (ed. Bright, 104), 2. 36 (ed. Bright, 105).
[111] Stead, *Divine Substance*, 260.
[112] *Ar*. 3. 65 (ed. Bright, 219); *Afros* 4 (*PG* 26. 1036). Cf. Stead, *Divine Substance*, 233.

rejected the primacy of the Father as ἀγένητος; as a Greek term did not replace the scriptural definition of God as Father in describing the divinity of the Son.[113] ἀγεννησία and γέννησις were the only proper distinctions within the Godhead, and preserved the Father as the ultimate source of divinity. On the other hand, as Stead points out, Athanasius retained a traditional, asymmetrical sense of divine nature: the Father was never said to be ὁμοούσιος with the Son; rather, the Son was from the Father or from the substance of the Father.[114] Even in his analogies to the fountain and stream or source and ray, Athanasius maintained the communication of being from the Father to the Son.[115] Athanasius therefore did not define ὁμοούσιος as a numerical identity, as did the later Cappadocians, but rather continued the more conservative sense of several sharing a common essence.[116] The Son as Image or Word received the full communication of the Father's essence, but the sense of the Father as source was not entirely diminished.

Not surprisingly, Athanasius carefully qualified his own use of will in reference to the Father and the Son: the Son was begotten from the essence of the Father as a Son, not by divine will as a work.[117] Thus, cautious of any arbitrary sense of will as unworthy of divine essence, Athanasius usually denied it a place in the generation of the Son. First, because the Son as divine must be eternal, generation by will, which implies temporality, is absurd; his existence is not akin to that of creatures in any way, for the creating Word of God is not literally the first-born creature.[118] Thus, the Son was always with God as the will of God, just as he was always Logos or Wisdom.[119] Generation was therefore shaped by his understanding of the divine nature as eternal and unchanging, as well as by the necessary divinity of the Son: 'As far then as the Son transcends the creature, by so

[113] *Ar.* 1. 30 (ed. Bright, 31–2), 1. 34 (ed. Bright, 35–6); *Decr.* 28–31 (*PG* 25. 468–75).

[114] *Syn.* 50 (*PG* 26. 781); Stead, *Divine Substance*, 261.

[115] *Dion.* 24 (*PG* 25. 516); *Decr.* 23–4 (*PG* 25. 456–7); Stead, *Divine Substance*, 263.

[116] Stead, *Divine substance*, 266.

[117] *Ar.* 1. 29 (ed. Bright, 30), 3. 62 (ed. Bright, 215).

[118] *Ar.* 2. 62 (ed. Bright, 132–3).

[119] *Ar.* 2. 31 (ed. Bright, 100), 2. 2 (ed. Bright, 70).

much does what is by nature transcend the will.'[120] The Son as begotten was therefore eternal and essential to divine being.

Yet, God was not constrained by necessity to generate the Son, since will and nature coincided in God:

For if they grant to God the willing [τὸ βούλεσθαι] about things which are not, why do they not recognize in God that which lies above will? For what he should be by nature, and that he is the Father of his own Word transcends will. (*Ar.* 2. 2 (ed. Bright, 70))

In an early work Athanasius had claimed that their existence was by necessity, but against these objections he later distinguished between deliberation (βούλησις) and desire (θέλημα); θέλησις was the interior expression of nature.[121] The desire of the Father for the Son therefore reflected the eternal communion of their shared nature rather than any contingent decision:

For as his own existence is by his desire [θελητής], so also the Son, being proper to his essence, is not without his desire [ἀθέλητός]. The Son is desired and loved by the Father. . . . For by that desire the Son is desired by the Father, and by it he himself loves, desires, and honours the Father. One is the will [θέλημα] from the Father in the Son, so that from this we may contemplate the Son in the Father and the Father in the Son. (*Ar.* 3. 66 (ed. Bright, 219))

Although Origen had a similar notion of the shared love and will between Father and Son, it was based on activity as much as common nature. For Athanasius, however, the idea of a common will followed on the idea of a common nature. Thus, the Son did not do the Father's will as somehow external to him; rather, he *was* the divine will, just as he was the Word or Wisdom. The Son revealed the divine will of the Father because it was his own, by both definition and nature.

However, this identification of the Son with divine will was mixed up with older language about the agency and obedience of the Son. In his early work *Against the Pagans*, Athanasius outlined a traditional apologetic of the cosmic agency of the Logos, emphasizing his power and will in the co-ordination of the universe; as in Origen and Eusebius, the Son remained the

[120] *Ar.* 3. 62 (ed. Bright, 215), 2. 31 (ed. Bright, 100).

[121] Contrast *Gent.* 47 (ed. Thomson, 130) with *Ar.* 3. 66 (ed. Bright, 219).

cosmological mediator between the Father and the world.[122] However, in later works the language of agency of the Son became appropriate only to the incarnational activities foreseen by God. Thus, some troublesome subordinationist texts usually applied to the essential relationship of the Father and the Son could be referred to the Incarnation.[123] God, foreseeing the fall of humanity, 'willed' the incarnation of the Son for human salvation.[124] In this sense the Son was the agent of the Father, although embodying the divine will in his actions; the action proved the identical nature: 'The Father is he who comes in the Son.'[125] However, in contrast to some earlier authors he saw the agency of the Son as a minor theme, one that was overshadowed by the effect of incarnation through his essential divinity, as in this exegesis of Hebrews 1: 2:

For if 'all things' were given to him, first he is other than all which he received. . . . For if he were one of all, then he were not heir of all, for everyone has received as the Father willed and gave. But receiving all, then he is other than them all and solely proper to the Father . . . So from saying 'have given', it shows that he is not the Father, but in saying 'so', it shows the Son's natural likeness. (*Ar.* 3. 36 (ed. Bright, 190–1))

In the same way in which he focused on the eternal relationship of the Father and the Son with regard to the Son's agency, he also distinguished language regarding the divine and the human nature. This took him away from a subordinationist exegesis; yet the hermeneutic of the divine Son also created problems in understanding Scripture and the authentic humanity of Christ:

When the Saviour used the words, which they cite as an excuse in their defence, 'Power is given to me' and 'Glorify your Son' . . . we understand all these passages in the same sense, that humanly on account of the body he says all this. For though he had no need, yet he is said to have received what he received humanly, so that as the Lord has received, and the gift is lodged in him, the grace may remain there. For

122 *Gent.* 44 (ed. Thomson, 120); *Inc.* 1 (*SC* 199. 259–60), 7 (*SC* 199. 286–8).

123 *Ar.* 2. 47 (ed. Bright, 116–17), 3. 26 (ed. Bright, 180–1).

124 *Ar.* 2. 51 (ed. Bright, 120), 2. 77 (ed. Bright, 147).

125 *Ar.* 1. 63 (ed. Bright, 66), 2. 76 (ed. Bright, 146), 2. 54–6 (ed. Bright, 123–7), 3. 31 (ed. Bright, 186); 3. 11 (ed. Bright, 165), 3. 35–6 (ed. Bright, 189–91).

what a mere man receives, he is liable to lose again, and this is shown in Adam. He received and lost it. So that grace may be irrevocable and may be kept sure for humanity, he himself appropriates the gift. (*Ar.* 3. 38 (ed. Bright, 193))

Athanasius's appeal to double exegesis allowed him to attribute passages concerning ignorance, suffering, or obedience to the human nature. Yet, as in the passage quoted above, the identity of Christ as divine Son tended to overshadow the authenticity of the human need or receptivity in the Incarnation. Unable to escape his own axiom of nature determining language, the categories of faithfulness, progress, and fear or ignorance tended to be shaped so as to be suitable for the divine Logos as subject of the action. Thus, the faithfulness of Christ reflected his humanity; but he was not faithful as we are, but deserved to receive faith.[126] Christ was not in need of deliverance, but brought deliverance.[127] Were he in need of grace, he could not bring it to others.[128] In spite of the principle of double exegesis, the Logos remained the primary subject.

These examples highlight one of the troublesome issues of his theology of incarnation. Clearly, Athanasius was not a docetic thinker. Unless the human nature were real, the enfleshment of the divine Son would not effect the necessary internal cleansing of the human body of the passions of birth and ignorance.[129] Yet the power of the divinity must be primary, in order to destroy the ravages of sin in fallen humanity: namely, mortality and corruptibility.

So while he himself being impassible by nature remains as he is, not hindered by these passions, but rather obliterating and destroying them, so that human beings, their passions transformed and abolished by the impassible become impassible and free. (*Ar.* 3. 34 (ed. Bright, 189))

Like Eusebius, Athanasius refers to Christ's humanity as *νάος* or *ὄργανον*, often giving it an instrumental sense.[130]

The actual structure and role of the humanity of Christ in his thought thus remain controversial, particularly in light of his relation to Apollinarius. He repeatedly stated that the Logos

[126] *Ar.* 2. 6–9 (ed. Bright, 74–7). [127] *Ar.* 2. 72 (ed. Bright, 142).

[128] *Ar.* 3. 38 (ed. Bright, 192–3), 3. 39 (ed. Bright, 193).

[129] *Ar.* 2. 68–9 (ed. Bright, 138–9), 3. 31–3 (ed. Bright, 185–8).

[130] See Kannengiesser, *Sur l'Incarnation*, (*SC* 199. 292–3).

took on the weaknesses of human nature; yet his actual exegesis tended to separate the Logos from the experiences of the flesh in order to defend his divinity. As the subject of the action, the Logos could not be ignorant. Thus, when Jesus asked a question, it could only be rhetorical, or he might pretend to be ignorant for the sake of human faith.[131] Because of the power of the Logos, he had to allow his body to be hungry or weep or be sick, or in fact die.

> He who did the works is the same as he who showed that his body was passible by allowing it to weep and hunger and to show other properties of a body. For by these means he made it known that being divine and impassible, he took passible flesh, yet from his works also showed himself to be the Word of God. (*Ar.* 3. 55 (ed. Bright, 208))

The Logos itself, therefore, was not polluted by the experiences.[132] Athanasius did not follow a full double exegesis such as would have allowed such experiences in human nature; rather, he applied them to the Logos as subject, as his predecessors had done. If they had compromised the divinity of the Son, his defence of the Son's divinity led him to twist the natural sense of scriptural passages in odd ways to explain the effect of human weakness on the divine.

The vehemence of his language suggests that his style of double exegesis may not have been as obvious to the fourth-century Church as he claimed. His exegesis was surrounded by appeal to authority (σκόπος) in scriptural interpretation and rhetorical appeals to common sense:

> For if we recognize what is proper to each, and see and understand that both these things are done by one, we believe correctly and shall never stray . . . but we having the scope of faith, recognize that what they interpret evily has a right interpretation. (*Ar.* 3. 35 (ed. Bright, 189f.))[133]

As shown in the structure of *Against the Arians*, interpretation of the scriptural texts regarding Jesus was highly controversial in the fourth century, ranging from the apologetic of Eusebius and

[131] *Ar.* 3. 37 (ed. Bright, 192), 3. 43 (ed. Bright, 197). See E. D. Moutsoulas, 'La Lettre d'Athanase d'Alexandrie à Epictète', in Kannengiesser (ed.), *Politique et théologie*, 313–33, and A. L. Petterson, 'The Questioning Jesus in Athanasius' *Contra Arianos III*', in Gregg (ed.), *Arianism*, 243–56.

[132] *Ar.* 3. 54–5 (ed. Bright, 206–9); *Inc.* 20–3 (*SC* 199. 336–50).

[133] *Ar.* 3. 58 (ed. Bright, 211), 3. 31 (ed. Bright, 185–6).

the dynamic monarchianism of Marcellus to the mediating subordinationism of Eunomius. Athanasius's scriptural hermeneutic of the divine Son was also problematic, because of his definitions of divinity as eternal and his view of humanity as essentially passive.

The presence or absence of a human soul in Christ in his early work remains problematic, although in his later writings, after the condemnation of Apollinarius, he seems to include it. Because of the criticism of Origen, most theologians in the early fourth century did not refer to Christ having a human soul, adopting instead a Logos–sarx Christology. In his early commentary on the Psalms, Athanasius followed Eusebius's exegesis, and occasionally referred to the soul or life of Christ, but to a lesser extent than Eusebius had done.[134] However, since the soul is not emphasized in his anthropology, its role in his Christology is less important. Richard has argued that this undermines his exegesis, since psychological weakness then has no clear subject.[135] His references to Christ as exemplar thus refer not to the strengthened human soul, but rather to the perfect endurance of the Logos:

> For as he taught us meekness from himself, saying 'Learn of me, for I am meek and lowly in heart', not that we may become equal to him, which is impossible, but that looking towards him, we may remain meek continually, willing that our good disposition towards each other should be true and firm and indissoluble, from himself taking the pattern. . . . For human conduct is very changeable, so one may look to what is unchangeable by nature and flee from what is bad and remodel himself on what is best. (*Ar.* 3. 20 (ed. Bright, 175))

This model of perfect virtue embodied by the incarnate Logos was of course traditional; but within his theological model there was little place for either a human *πρόσωπον* or a relationship of the Father to the Son that would reflect biblical accounts of obedience or agency.

Will was no longer a reflection of redemptive power or movement, having a lesser place in Athanasian Christology, because of being profoundly linked to mutable nature. Against the Arian description of Jesus as *αὐτεξούσιος*, Athanasius denied that it is proper or consistent with an affirmation of divinity. How could

[134] Stead, 'Scriptures and Soul of Christ'. [135] Richard, 'Saint Athanase'.

the Son advance and reveal the Father? Or if he acquired an ἄτρεπτος nature by will, when did he reveal the Father?[136] Moral action was impossible, for this would compromise the unchanging divinity of the Son.[137] Because revelation and physical transformation lie at the heart of Athanasian Christology, an obedient human will, if implying a created nature, was largely extraneous. According to Athanasius, the Arians repeatedly asked 'Has he free will or has he not? Is he good from choice according to free will, and can he, if he will, alter, being of alterable nature?'[138] Gregg and Groh have argued that this question reflects the Arian concern for a Christ who would be an authentic model of salvation for free, willed creatures.[139] Given Athanasius's definition of unchanging divine nature and the unpredictability of human will, his response to this question and his exegesis of Psalm 45: 7–8 are clear:

> Because of this was a need of one unchangeable, that humanity might have the immutability of the Word as an image and type for virtue. . . . For since the first man Adam changed and 'through sin death came into the world', therefore the second Adam needed to be unchangeable . . . so the Lord who ever is in nature unchangeable, 'loving righteousness and hating iniquity', should be anointed and sent, that he being and remaining the same, by taking changeable flesh . . . might secure its freedom. (*Ar.* 11. 51 (ed. Bright, 53–5))

The second Adam whom Athanasius described was therefore unchangeable and just by nature, because he was the incarnate Logos. In spite of his double exegesis, his view of human will and Logos–sarx Christological structure did not allow for the human perfection of a second Adam as in Irenaeus or Origen. To portray such activity inclined one towards adoptionism. For just as God was righteous by nature and hence above choice, so was the incarnate Son.[140]

The pattern, or model, which the Son brought therefore revealed the power of divine nature. As for Origen and Eusebius, αὐτεξούσιος was not an appropriate characterization of the divine Logos. However, the identification of the Son with the Father, while enhancing incarnational soteriology, sacrificed some

[136] *Ar.* 3. 51 (ed. Bright, 204).
[137] *Ar.* 3. 11 (ed. Bright, 164–5).
[138] *Ar.* 1. 35 (ed. Bright, 36).
[139] Gregg and Groh, *Early Arianism* 161f.
[140] *Ar.* 1. 51–2 (ed. Bright, 53–5).

of the richness of earlier exegesis. Origen contrasted the will of the Father with that of the Son, whereas Eusebius described the obedience of the human *πρόσωπον* of the Logos; but Athanasius more consistently celebrated the soteriological effect of divine power over human flesh:

For he willed what he deprecated, for this he had come; but his was the willing, for terror belonged to the flesh. Thus as human he speaks in this voice, yet both were said by one to show that he was God, willing in himself, but when he became human, he had a flesh that was in terror. For the sake of this flesh he combined his own will with human weakness, that destroying his suffering he might in turn make people undaunted in face of death. . . . But from this the most enduring purpose and courage of the holy martyrs is shown that God was not in terror, but the Saviour took away our terror. (*Ar.* 3. 57 (ed. Bright, 209))

In many ways this passage reflects the stalwart Son of Origen's and Eusebius's accounts. In all three, the Logos remains the subject of the action. However, for Origen and Eusebius, a significant aspect is the obedience of the Son to the plan of salvation in accepting death and the revelation of divine love by this self-sacrifice; the obedience of the incarnate subject is somehow consistent with the agency of the Logos in regard to his relationship with the Father and his self-offering in the Incarnation. For Athanasius, the transforming nature of divinity was theologically primary; hence the prayer in Gethsemane was split into two, to show the weakness of humanity and the power of the Logos. The martyrs no longer represent human endurance perfected through divine power, but rather the decisive transformation of human nature through the Incarnation.

Nevertheless, Athanasius did not discard the language of will entirely in regard to the Logos. In spiritual and pastoral letters he referred to Christ as an exemplar:

Thus, even our Lord and Saviour Jesus Christ comes before us, when he would show humanity how to suffer, who when he was smitten bore it patiently . . . and at last was willingly led to death that we might behold in him the image of all that is virtuous and immortal, and that we, conducting ourselves after these examples, might truly tread on . . . all the power of the enemy. (Letter 10 (tr. Robertson, 530))[141]

He describes here the heroic activity of the Word in becoming

[141] On Christ as exemplar, see also *Ep Marc.* 13 (*PG* 27. 25); and above, n. 92.

incarnate and modelling the power of discipleship. In many ways this passage echoes earlier accounts of the agency of the Logos; however, within his theological model, the emphasis now lies on the perfection of power in the imitation of the 'immortal' virtue of Christ. Moral freedom is not denied, but it is no longer primary, and therefore does not require a model in Christ.

Soteriology in Athanasius thus centres on the communication of the qualities of transcendent divine nature to mutable humanity through the incarnation of the Son. The Word by taking flesh was united to humanity in order to effect a whole transformation for its redemption, rather than affecting human behaviour only through teaching or example.[142] The body of Christ thus allows other human bodies to be temples of God, because it has stabilized the physical weakness which is the product of sin. Sin is associated with mutability and death, and thus humanity is delivered from contingency into absolute unity with God; thus redeemed humanity is higher than Adam. The key to this stability is the revelation of God in his image, the Son in the Incarnation. The re-creation is therefore intellectual as much as physical: the mind and the body have both been transformed through the Incarnation. Since disobedience was a consequence of the weakness of the body and the disorientation of the mind, the will is therefore to be strengthened through the transformation of the whole person.

This focus on both intellectual and material divinization is based on Athanasius's Christian Platonism and passionate ecclesiatical asceticism. Given a wholly transcendent God, only a fully divine, incarnate Logos can communicate knowledge of his nature. Although this cut the contemporary knots of epistemology and mortality, it meant that Athanasius's Christology had to ignore several traditional streams of biblical and Logos Christology. To affirm the divinity of the Son as eternal and essential to the Father, Athanasius was forced to reduce the sense of agency found in Christian cosmology since the Apologists. Indeed, such language seemed irrevocably adoptionist to him. This lack of attention to Christ's will and a general pessimism regarding the human will seem in stark contrast to earlier Christian teaching. Moral action no longer has a metaphysical dimension. Yet, at heart, Athanasius's theology remains as much

[142] Ricken, 'Rezeption', 348–9; Young, *Nicaea to Chalcedon*, 69.

Logocentric as Christocentric. In many passages the Son remains the intellectual and cosmological mediator between the transcendent Father and creation: 'If not created in Him, we have Him not in ourselves but externally, as for instance as receiving instruction from him as from a teacher.'[143]

Historically, one must recognize the profound interaction between contemporary Platonism and Athanasius's theology, rather than appeal to an essential biblicism or kerygma in his defence of the divine incarnation. What separated Athanasius from his predecessors was a new religious context, as well as theological controversy, which led him to emphasize certain Christian themes at the expense of others. Ironically, in a setting of increasing social legitimacy for Christianity, this incarnational view cast Christian identity in starker form over against the larger world or its history. Athanasius's claim that God incarnate dwelt in the suffering Christ is wholly unplatonic; yet it is fuelled by contemporary assumptions about divine transcendence, noetic divinization, and material frailty, so that the role of the Son remains profoundly cosmological in reuniting and redeeming the fallen creation. However, as reflected in his theological model, cosmology has shifted. Soteriology is no longer focused on deliverance from fatalism through obedience and instruction, but deliverance from changeability through physical transformation. If such categories reflect an unreal antithesis philosophically, they were clearly in line with contemporary asceticism's rejection of conventional social and perhaps ecclesiastical identity. By this synthesis of ecclesiastical identity and ascetic power through a singular divine presence in the Incarnation, Athanasius sharply defined the Christian community over against Graeco-Roman culture, and sought to incorporate the scattered ascetics into the mainstream. The fourth-century heavens remained open, but the process was now weighted through a corporate, ecclesiastical conduit, rather than a multiplicity of individual journeys. Like those of his predecessors, Athanasius's compelling theological vision derived strength from a Christian transposition of contemporary trends to underline soteriology and cosmology, with Christ as the decisive pivot.

143 *Ar.* 2. 56 (ed. Bright, 126); cf. *Ep. Marc.* 13 (*PG* 27. 25).

CONCLUSION

> You know the Greeks aren't broad church clergymen. They really aren't, in spite of much conflicting evidence.
>
> (E. M. Forster, *The Longest Journey*)

Ancient reflection on the nature and activity of Christ necessarily depended on contemporary philosophical definitions of divinity and humanity; but the biblical accounts of Jesus's actions and relationship with the Father profoundly shaped the resulting theological models. Logos Christology was not a passive product of cosmological categories, but a creative, passionate model of an emerging world-view. In their apologetic and polemical presentations of Christology, Origen, Eusebius, and Athanasius each borrowed and modified common formulas of Late Antiquity to express particular theological concerns. These concerns were not only individual beliefs, however; they also represented the spiritual and social realities of the communities which shaped the focus and language of each man: small urban study groups, public civic apologetics, and the emerging ascetical church.

The emphasis on the subordinate agency of the Son which lasted until the fifth century did not necessarily reflect a philosophical dominance or the lack of a coherent soteriology in early Christianity. As shown in this and other recent studies, the controversies of the fourth century exposed the variety of conflicting theological models within Christianity, rather than a simple contrast between biblical, soteriological perspectives and cosmological, philosophical views. However, these traditional typologies remain difficult to shed. Thus, A. H. Armstrong contrasts the natural theology of the Greeks, who found meaning in the cosmos, with Christian belief in revelation only in the church through the Incarnation.[1] J. Lienhard has constructed a new typology for the fourth century, which attempts to contrast two means of salvation rather than cosmology and soteriology: 'mia

[1] Armstrong, 'Man in the Cosmos'.

hypostatic' and 'duo hypostatic'. Although he describes the biblical influences on subordination sympathetically, this soteriology is ultimately inadequate in his view because in place of divinization it 'offers a concept of salvation that is really no more than moralism'.[2] R. Williams recently claimed that a mediator could not bring *theosis*.[3] Such typologies preserve the inherited distinctions between Hellenism and Christianity, orthodoxy and heresy, but continue to obscure the actual history of early Christianity. As seen above, ancient Christians appealed to the evidence of the cosmos in defence of the doctrine of creation and providence, defended free will as critical to transforming union with God, and believed that divinization did indeed happen through a mediator.

Defining the subordination of the Son as an essentially soteriological error unfortunately forced early reflection into later Trinitarian or Christological models. It was not that inaccurate ontological considerations were put into an orthodox model, but rather, that different, contemporary issues of fatalism, equally soteriological, produced a different theological model of divine activity and human response. The cosmic and historical agency of Christ affirmed both God's continual activity within the world and the authenticity of human action. Analytically, the model of obedience presented in the Incarnation was of course one of exemplar rather than actual example.[4] Nevertheless, in the light of contemporary fatalism, it functioned devotionally to encourage Christian conversion and martyrdom, social autonomy, and spiritual progress. Like the equally problematic, religiously powerful affirmations of his suffering, the obedience of the divine Son was a sign of his unity with humanity and an affirmation of the process whereby God had chosen to redeem the world. In labelling these early models 'merely moral', 'external', or dominated by a classical idea of παίδεια, scholars have ignored both the biblical influences within the model of agency and obedience and the radical claims being made about divine power and human transformation.

This is not so much to minimize the distinctions between Christian and Greek world-views as to move away from static

[2] Lienhard, '"Arian" Controversy', 430–1. [3] Williams, *Arius*, 241.

[4] For contrast of 'exemplar' with 'example', see Stead, 'Freedom of Will', 255.

definitions of each, in order to understand the creativity and religious motivations of earlier authors in their cosmological reflections. The Hellenization of Christianity continues to be a critical issue in modern theology, and the use of ancient philosophy needs to be understood within its historical and religious context.[5] As M. Douglas has commented in regard to anthropological studies, religious language itself does not explain everything; rather, religion as a world-view must be explained.[6] Categories such as incarnation, soteriology, and transcendence do not immediately reveal their theological function without attention to their religious and social contexts. The multiplicity of models within the ancient Alexandrian theological tradition reflected in part concrete distinctions among early Christian communities. Origen, Eusebius, and Athanasius all affirmed the pre-existent divinity of the Son, his incarnation as a fully human person, and the inability of individuals to save themselves apart from the grace of God; and all assumed a Platonic ontology. However, the pressures of fatalism or transcendence pushed Christological models and philosophical synthesis in different directions.

In conclusion, I want to summarize some 'correlations' which seem to exist between the cosmological models and historical communities of these three authors.[7] In a century and a half of changing social context and religious identity, each author reflected both religious optimism and social alienation. In the intellectual and social ferment of Alexandria, albeit in a largely stable empire, Origen assumed a spiritual mobility which would allow the individual ascent and transformation of each soul under the direct tutelage of the Logos; the ἐπίνοιαι in some sense expressed the social and spiritual multiplicity of the early urban Christian communities engaged with a new sacred text in seeking union with a new God. Shaped by the contemporary issues of freedom and fatalism, especially as translated into Christian form by the Gnostics, Origen's cosmology defended the

[5] Examples of recent discussions may be found in R. Radford Ruether, *To Change the World* (New York, 1981); R. J. Schreiter, *Constructing Local Theologies* (Maryknoll, NY, 1986); R. E. Hood, *Must God Remain Greek?* (Philadelphia, 1990).

[6] Douglas, *How Institutions Think*, 112.

[7] 'Correlation' is the term used by Meeks, *First Urban Christians*, 191.

self-determination of the individual mind and soul; even the body, though a hindrance, was essentially fluid and capable of being transformed in accordance with spiritual progress towards union with God. As a teacher and exemplar, Christ's agency was a critical hinge between divine goodness and human freedom; the accommodation of the Logos through the ἐπίνοιαι affirmed the multiplicity of spiritual journeys towards a final unity. The perfect obedience of Jesus was an important model for both martyr and convert struggling to sustain a new identity. Although the action of Christ's human soul was ideal because of its unique adhesion to the Logos, imitation of this volitional stability in relation to God was the goal of each Christian. The active will of a good God therefore ensured an optimistic, individualized, open cosmology. Significantly, even in the final union with God, individual distinctions remained.

Eusebius's broad, sweeping vision of the public triumph and dominance of Christianity rested on the unmistakable historical signs of divine power and social progress. The acts of Jesus, whose power was continued in the church, made sense of all historical human hope and endeavour from Moses and Plato to Constantine. Amidst the increasing sense of transcendence and the social disorders of the early fourth century, Eusebius offered the Christian community as an assurance of God's utter control and intimate presence in history. By contrast with Origen, his sense of progress and spirituality were therefore more social and communal, if equally optimistic. He too believed that the body could be controlled by the individual will through the teaching and power of the incarnate Logos; but this transformation was most clearly seen in the progress, punishment and repentance, and new flourishing of the church. If the bodies and souls of the poor, the uneducated, women, and even philosophers could be divinized and moulded into one community, so could the entire known world in accordance with divine will. Rather like the incarnation of the Logos, the Christian church appeared to be a historical aberration, but in fact was the culmination of salvation history under the patient power of divine will. The evidential sense of Eusebius's Christology is thus linked with the collective sense of his soteriology. The cosmic and historical agency of the Logos unified and assured salvation history; the focus on the obedience of the Son is less as an individual exemplar than as an affirmation of continuous, saving providence within the structures of existence. Like Athanasius, Eusebius proposed an ecclesiastical Christology which celebrated the Incarnation as an objective act which constituted the beginning of a new order of history and existence.

Coming of age in an era of social legitimacy and imperial patronage of the Christian church, Athanasius's presentation of the Incarnation was anchored in the emerging identity of the Egyptian church. His defence of the fully divine Son's assumption of human flesh to transform material life reflects both ascetic separation from the surrounding world and contemporary problems of divine transcendence. The body has become a stubborn sign of human sin and alienation, particularly in a corporate sense. In contrast to his predecessors, Athanasius did not primarily defend individual progress or celebrate the evidence of social transformation, but rather insisted on the necessary transformation of all flesh through the Word. Like ascetics, everyday Christians were therefore set apart from the larger society, required to live as angels in a disintegrating world. Activity no longer transformed and redeemed nature; rather, the redemption of nature stabilized human activity. The language of moral striving thus became largely detached from cosmological change and divinization. Not surprisingly, the agency of the Logos was less important cosmologically or historically for Athanasius. Instead, the assertion of the transforming presence of the divine nature in flesh became the key to religious devotion and social transformation. The incarnation of the Logos embodied the corporate, stable uniqueness of the ascetic church in late Roman society. The singular, arresting presence of God in Christ, now set more starkly over against history, society, and the Graeco-Roman tradition, became the locus of salvation.

BIBLIOGRAPHY

PRIMARY TEXTS AND TRANSLATIONS

Athanasius

PG 25: *De decretis.*
PG 26: *Epistolae de Synodis Arimini in Italia et Seleuciae in Isauria celebratis; Epistola ad Afros; Epistolae ad Serapionem; Epistola ad Epictetum; Epistolae festales.*
PG 27: *Epistola ad Marcellinum.*

BRIGHT, W. (ed.), *The Orations of St Athanasius against the Arians according to the Benedictine Text* (2nd edn., Oxford, 1884).

KANNENGIESSER, C. (ed. and tr.), *Sur l'Incarnation du Verbe*, *SC* 199 (Paris, 1973).

MEIJERING, E. P. (ed. and tr.), *Athanasius. Contra gentes. Introduction, Translation, and Commentary*, Philosophia Patrum, 7 (Leiden, 1984).

OPITZ, H. G. (ed.), *Athanasius Werke*, 3. 1. *Urkunden zur Geschichte des arianischen Streites 318–328* (Berlin, 1935).

ROBERTSON, A. (ed. and tr.), *Saint Athanasius. Select Works and Letters*, *NF* 4 (repr. Grand Rapids, 1978).

THOMSON, R. (ed. and tr.), *Contra Gentes and De Incarnatione*, Oxford Early Christian Texts (Oxford, 1971).

Eusebius

GCS 7: *De laudibus Constantini*, ed. I. A. Heikel.
GCS 11. 2: *Theophania*, ed. H. Gressmann.
GCS 14: *De ecclesiastica theologia; Contra Marcellum*, ed. E. Klostermann.
GCS 23: *Demonstratio evangelica*, ed. I. A. Heikel.
GCS 43: *Praeparatio evangelica*, ed. K. Mras.

CONYBEARE, F. C. (tr.), 'Against Hierocles', in *The Life of Apollonius of Tyana. The Epistles of Apollonius and the Treatise of Eusebius*, Loeb Classical Library (London, 1950).

FERRAR, W. J. (ed. and tr.), *The Proof of the Gospel* (London, 1920; repr. Grand Rapids, 1981).

GIFFORD, E. H. (ed. and tr.), *Evangelicae praeparationis* (4 vols., Oxford, 1903).

LAKE, K. (ed.), *The Ecclesiastical History*, Loeb Classical Library (Cambridge, Mass., and London, 1926; repr. 1975).

Origen

GCS 2: *Exhortatio ad martyrium*, ed. P. Koetschau.
GCS 3: *De Oratione*, ed. P. Koetschau.
GCS 29: *Homiliae in Genesim, Exodum, Leviticum*, ed. W. A. Baehrens.
GCS 33: *Homiliae in 1 Reges; Commentaria in Canticum Canticorum; Homiliae in Isaiam, Homiliae in Ieremiam* (Latin), *Homiliae in Ezechelem*, ed. W. A. Baehrens.
GCS 40: *Commentaria in Matthaeum*, ed. E. Klostermann.
GCS 41: *Fragmenta in Matthaeum*, ed. E. Klostermann.
PG 12: *Commentaria in Genesium.*
PG 14: *Commentaria in epistolam ad Romanos.*

BLANC, C. (ed. and tr.), *Commentaire sur Saint Jean*, *SC* 120, 157, 222, 290 (Paris, 1966–81).

BORRETT, M. (ed. and tr.), *Contre Celse*, *SC* 132, 136, 147, 150 (Paris, 1967–9).

——(ed. and tr.), *Homélies sur le Lévitique*, *SC* 286–7 (Paris, 1981).

BROOKE, A. E. (ed.), *The Commentary of Origen on S. John's Gospel* (2 vols., Cambridge, 1896).

BUTTERWORTH, G. W., *On First Principles* (London, 1936; repr. New York, 1973).

CHADWICK, H. (ed. and tr.), *Contra Celsum* (Cambridge, 1953; repr. 1980).

CROUZEL, H., and SIMONETTI, M. (eds. and trs.), *Traité des Principes*, *SC* 252–3, 268–9, 312 (Paris, 1978–84).

CROUZEL, H., FOURNIER, F., and PERICHON, P. (eds. and trs.), *Homélies sur Saint Luc*, *SC* 87 (Paris, 1962).

GREGG, J. A. F. (ed.), 'The Commentary of Origen upon the Epistle to the Ephesians', *JTS* 3 (1901–2), 233–44, 398–420, 554–76.

JENKINS, C. (ed.), 'Origen on I Corinthians, IV', *JTS* 10 (1908–9), 29–51.

LAWSON, R. P. (tr.), *The Song of Songs. Commentary and Homilies*, Ancient Christian Writers, 26 (London, 1957).

NAUTIN, P. (ed. and tr.), *Homélies sur Jérémie*, *SC* 232, 238 (Paris, 1976–7).

ROBINSON, J. (ed.), *Philocalia* (Cambridge, 1893).

SCHERER, J. (ed.), *Entretien d'Origène avec Heraclide* (Cairo, 1949).

——*Le Commentaire d'Origène sur Rom. III.5–V.7 d'après les extraits du papyrus No. 88748 du Caire et les fragments de la Philocalie et du Vaticanus Gr. 762* (Cairo, 1957).

Other Sources

The Acts of the Christian Martyrs, ed. H. Musurillo, Oxford Early Christian Texts (Oxford, 1972).

Albinus, *Épitomé*, ed. P. Louis, Nouvelle collection de textes et documents publiée sous le patronage de l'Association Guillaume Budé (Paris, 1945).

Alexander of Lycopolis, *An Alexandrian Platonist against Dualism. Alexander of Lycopolis' Treatise 'Critique of the Doctrines of the Manichaeus'*, ed. and tr. P. W. van der Horst and J. Mansfeld (Leiden, 1974).

The Apostolic Fathers, ed. and tr. K. Lake, Loeb Classical Library (Cambridge, Mass., and London, 1912; repr. 1975).

Athanassakis, A. (ed. and tr.), *The Life of Pachomius (Vita Prima Graeco)*, Texts and Translations, 7, Early Christian Literature Series, 2 (Missoula, Mont., 1975).

Atticus, *Fragments*, ed. and tr. É. des Places, Collection des universités de France publiée sous le patronage de l'Association Guillaume Budé (Paris, 1977).

Bayer, K. (ed.), *Justin. Philosoph und Martyrer. Die Erste Apologie* (Munich, 1966).

Gregory Thaumaturgus, *Remerciement à Origène. La lettre d'Origène à Grégoire*, ed. and tr. H. Crouzel, *SC* 148 (Paris, 1969).

Irenaeus, *Contre les hérésies*, ed. and tr. A. Rousseau and L. Doutreleau, *SC* 100, 152–3, 210–11, 263–4, 293–4 (Paris, 1965–82).

Justin, *Die altesten Apologeten*, ed. F. Goodspeed (Göttingen, 1914).

Mani, *Kephalaia*, ed. H. Ibscher (Stuttgart, 1940).

Numenius, *Fragments*, ed. and tr. É. des Places, Collection des universités de France publiée sous le patronage de l'Association Guillaume Budé (Paris, 1973).

Plotinus, *Opera*, ed. P. Henry and H.-R. Schwyzer, Museum Lessianum, Series Philosophia, 24 (Paris and Brussels, 1951–73).

Robinson, J. M. (ed.), *The Nag Hammadi Library in English* (3rd rev. edn., San Francisco, 1988).

Todd, R. (ed. and tr.), *Alexander of Aphrodisias on Stoic Physics. A study of* De Mixtione *with Text, Translation, and Commentary* (Leiden, 1976).

Walzer, R. (ed.), *Galen on Christians and Jews* (London, 1949).

Ward, B. (tr.), *The Sayings of the Desert Fathers. The Alphabet Collection* (Kalamazoo, Mich., 1975).

Winden, J. C. M. van (ed.), *An Early Christian Philosopher. Justin Martyr's Dialogue with Trypho*, Philosophia Patrum, 1 (Leiden, 1971).

SECONDARY MATERIALS

AEBY, G., *Les Missions divines de Saint Justin à Origène* (Fribourg, 1958).

ALAND, B., 'Erwählungstheologie und menschenklassen Lehre. Die Theologie des Herakleon als Schlüssel zum Verständnis der christlichen Gnosis?', in M. Krause, J. M. Robinson, and F. Wisse (eds.), *Gnosis and Gnosticism*, Nag Hammadi Studies, 8 (Leiden, 1977), 148–81.

ALCAIN, J. A., *Cautiverio y redencion del hombre en Origenes* (Bilbao, 1973).

AMAND DE MENDIETA, E. A., *Fatalisme et liberté dans l'antiquité grecque*, Université de Louvain, recueil de travaux d'histoire et de philosophie, 3e ser., fasc. 19 (Louvain, 1945).

ANDRESEN, C., 'Justin und der mittlere Platonismus', *ZNW* 44 (1952–3), 157–95.

ARMSTRONG, A. H., *The Architecture of the Intelligible Universe* (Amsterdam, 1967).

——'Gnosis and Greek Philosophy', in B. Aland (ed.), *Gnosis. Festschrift für Hans Jonas* (Göttingen, 1978), 87–124.

——'Man in the Cosmos. A Study of some Differences between Pagan Neoplatonism and Christianity', in W. den Boer, P. G. van der Nat, and J. C. M. van Winden (eds.), *Romanitas et Christianitas. Studia J. H. Waszink* (Amsterdam and London, 1973), 5–14.

——'The Self-Definition of Christianity in Relation to Later Platonism', in E. P. Sanders (ed.), *Jewish and Christian Self-Definition*, Vol. 1: *The Shaping of Christianity in the Second and Third Centuries* (London, 1980), 74–99.

——'Two Views of Freedom. A Christian Objection in Plotinus *Enneads* VI. 8 (39)7, 11–15?', *SP* 17.1 (1982), 397–406.

——(ed.), *Classical Mediterranean Spirituality* (New York, 1986).

——and MARKUS, R.A., *Christian Faith and Greek Philosophy* (New York, 1960).

BARNARD, L., 'Athanase, Constantin, et Constance', in C. Kannengiesser (ed.), *Politique et théologie chez Athanase d'Alexandrie*, Théologie historique, 27 (Paris, 1974), 127–43.

——*Justin Martyr. His Life and Thought* (Cambridge, 1967).

BARNES, T. D., *Constantine and Eusebius* (Cambridge, Mass., 1981).

——'The Emperor Constantine's Good Friday Sermon', *JTS*, NS 27 (1976), 414–23.

——'Porphyry *Against the Christians*: Date and the Attribution of Fragments', *JTS*, NS 24 (1973), 424–42.

BARR, J., 'Athens or Jerusalem? The Question of Distinctiveness', in idem, *Old and New in Interpretation* (London, 1966), 34–64.

BENZ, E. W., *Marius Victorinus und die Entwicklung der abendlandischen Willensmetaphysik*, Forschung zur Kirchen und Geistgeschichte, 1 (Stuttgart, 1932).

BERCHMAN, R. M., *From Philo to Origen* (Chico, Calif., 1984).

BERKHOF, H., *Die Theologie des Eusebius von Caesarea* (Amsterdam, 1939).

BERNER, U., *Origenes* (Darmstadt, 1981).

BETZ, H. D., 'Cosmogony and Ethics in the Sermon on the Mount', in R. W. Lovin and F. E. Reynolds (eds.), *Cosmogony and Ethical Order. New Studies in Comparative Ethics* (Chicago, 1985), 158–76.

BIANCHI, U., 'Anthropologie et conception du mal. Les sources de l'exégèse gnostique', *VC* 25 (1971), 197–204.

BIENERT, W. A., *Dionysius von Alexandrien. Zur Frage des Origenismus im drittem Jahrhundert* (Berlin, 1978).

BLANC, C., 'L'Attitude d'Origène a l'égard du corps et de la chair', *SP* 17.2 (1982), 843–58.

——'Dieu est *pneuma*. Le Sens de cette expression d' après Origène', *SP* 16.2 (1985), 224–41.

BOSTOCK, D., 'Quality and Corporeity in Origen', in H. Crouzel and A. Quacquarelli (eds.), *Origeniana Secunda*, Quaderni di 'Vetera Christianorum', 15 (Rome, 1980), 323–37.

BOYD, W. J. P., 'Origen on Pharaoh's Hardened Heart. A Study of the Justification and Election in St. Paul and Origen', *SP* 7 (1963), 434–42.

BRENNAN, B., 'Athanasius' *Vita Antonii*: A Sociological Interpretation', *VC* 39 (1985), 209–27.

BROWN, P. R. L., *The Body and Society* (New York, 1988).

——'Late Antiquity', in P. Veyne (ed.), *A History of Private Life*, vol. 1 (Cambridge, Mass., 1987), 235–311.

——*The Making of Late Antiquity* (Cambridge, 1978).

——*The Philosopher and Society in Late Antiquity*, ed. E. Hobbs and W. Wuellner, CHS 34, (Berkeley, Calif., 1980).

——*A Social Context to the Religious Crisis of The Third Century A.D.*, ed. W. Wuellner, CHS 14, (Berkeley, Calif., 1975).

——*The World of Late Antiquity* (London, 1971).

BROWN, R. F., 'On the Necessary Imperfection of Creation: Irenaeus *Adversus Haereses* IV. 38', *Scottish Journal of Theology*, 28 (1975), 17–25.

CARPENTER, H. J., 'Popular Christianity and the Theologians in the Early Centuries', *JTS*, NS 14 (1963), 294–310.

CHADWICK, H., 'Origen, Celsus and the Resurrection of the Body', *HTR* 41 (1948), 83–102.

Chadwick, H., *The Role of the Christian Bishop in Ancient Society*, ed. E. Hobbs and W. Wuellner, CHS 35 (Berkeley, Calif., 1980).

Chesnut, G. F., *The First Christian Histories: Eusebius, Socrates, Sozomen, Theodoret and Evagrius* (2nd rev. edn., Macon, Ga., 1986).

Cornélis, H., 'Les Fondements cosmologiques de l'eschatologie d'Origène', *Revue des sciences philosophiques et théologiques*, 43 (1959), 32–80, 201–47.

Countryman, L. W., 'The Intellectual Role of the Early Catholic Episcopate', *CH* 48 (1979), 261–8.

Crouzel, H., 'L'Anthropologie d'Origène dans la perspective du combat spirituel', *Revue d'ascétique et de mystique*, 31 (1955), 364–85.

—— 'L'Anthropologie d'Origène: de l'arché au telos', in U. Bianchi and H. Crouzel (eds.), *Arché e Telos. L'antropologia di Origène e di Gregorio di Nissa. Analisi storico-religiosa*, Studia Patristica Mediolanensia, 12 (Milan, 1981), 36–43.

—— 'Les Critiques addressés par Méthode et ses contemporains à la doctrine origénienne du corps ressuscité', *Gregorianum*, 53 (1972), 679–716.

—— 'The Literature on Origen 1970–1988', *TS* 49 (1988), 499–516.

—— *Origen*, tr. A. S. Worrall (San Francisco, 1989).

—— *Origène et la philosophie* (Paris, 1962).

—— 'Les Personnes de la Trinité sont-elles de puissance inégale selon Origène, Peri Archon 1, 3, 5–8', *Gregorianum*, 57 (1976), 109–25.

—— *Théologie de l'image de Dieu chez Origène* (Paris, 1956).

Daly, R. J., 'Origen Studies and Pierre Nautin's *Origène*, *TS* 39 (1978), 508–19.

Dempf, A., 'Der Platonismus des Eusebius, Victorinus und pseudo-Dionysius', in *Sitzungsberichte. Bayerische Akademie der Wissenschaften*, Philosophisch-historische Klasse, Jahrgang 1962, no. 3, 1–18.

Dewart, J. M., '"Moral Union" in Christology before Nestorius', *Laval théologique philosophique*, 32 (1976), 283–99.

Dihle, A., *The Theory of the Will in Classical Antiquity* (Berkeley, Calif., 1982).

Dillon, J., 'The Descent of the Soul in Middle Platonic and Gnostic Theory', in B. Layton (ed.), *The Rediscovery of Gnosticism. Proceedings of the International Conference on Gnosticism at Yale*, vol. 1 (Leiden, 1980), 357–64.

—— *The Middle Platonists: A Study of Platonism 80 B.C. to A.D. 220* (London, 1977).

—— 'Plotinus, Philo, and Origen on the Grades of Virtue', in H. D. Blume and F. Mann (eds.), *Platonismus und Christentum. Festschrift für Heinrich Dörrie*, Jahrbuch für Antike und Christentum, suppl. vol. 10 (1983), 92–105.

——'Plutarch and Second Century Platonism', in A. H. Armstrong (ed.), *Classical Mediterranean Spirituality* (New York, 1986), 214–29.

DODDS, E. R., *Pagan and Christian in an Age of Anxiety* (Cambridge, 1965).

DÖRRIE, H., 'Divers aspects de la cosmologie de 70 av. J.-C. à 20 ap. J.-C.', *Revue de théologie et de philosophie*, 3.22 (1972), 400–5.

——'*Formulae Analogiae*: An Exploration of a Theme in Hellenistic and Imperial Platonism', in H. J. Blumenthal and R. A. Markus (eds.), *Neoplatonism and Early Christian Thought* (London, 1981), 33–49.

——'Logos-Religion? oder Nous-Theologie? Die hauptsächlichen Aspekte des kaiserzeitlichen Platonismus', in J. Mansfeld and L. M. de Rijk (eds.), *Kephalaion. Studies in Greek Philosophy and its Continuation Offered to Professor C. J. de Vogel*, Philosophical Texts and Studies, 23 (Assen, 1975), 115–36.

——'Was ist "spätantiker Platonismus"? Uberlegungen zur Grenzziehung zwischen Platonismus und Christentum', *Theologische Rundschau*, NS 36 (1971), 285–302.

DOUGLAS, M., *How Institutions Think* (Syracuse, NY, 1986).

——*Natural Symbols. Explorations in Cosmology* (London, 1978).

DRAKE, H. A., *In Praise of Constantine. A Historical Study and New Translation of Eusebius' Tricennial Orations*, University of California Publications: Classical Studies, 15 (Berkeley, Calif. 1976).

DREWERY, B., *Origen and the Doctrine of Grace* (London, 1960).

DUPUIS, J., *'L'Esprit de l'homme'. Étude sur l'anthropologie religieuse d'Origène*, Museum Lessianum, section théologique, 62 (Bruges, 1967).

EICHINGER, M., *Die Verklärung Christi bei Origenes. Die Bedeutung des Menschen Jesus in seiner Christologie*, Wiener Beitrage zur Theologie, 23 (Vienna, 1969).

FORSTER, E. F., *The Longest Journey* (New York, 1962).

FOUCAULT, M., *The Archaeology of Knowledge and the Discourse on Language*, tr. A. M. Sheridan Smith (New York, 1972).

FOWDEN, G., 'The Pagan Holy Man in Late Antique Society', *Journal of Hellenic Studies*, 52 (1982), 33–59.

FOX, R. Lane, *Pagans and Christians* (New York, 1986).

FREND, W. H. C., 'Athanasius as an Egyptian Christian Leader in the Fourth Century', in idem, *Religion Popular and Unpopular in the Early Christian Centuries* (London, 1976), 20–37.

GEERTZ, C., 'Ethos, World View and the Analysis of Sacred Symbols', in idem, *The Interpretation of Cultures* (New York, 1973), 126–41.

——Religion as a Cultural System', in M. Banton (ed.), *Anthropological Approaches to the Study of Religion* (London, 1966), 1–45.

GORDON, R., 'Fear of Freedom? Selective Continuity in Religion during the Hellenistic Period', *Didaskalos*, 4 (1972), 48–60.

GRAESAR, A., *Plotinus and the Stoics. A Preliminary Study*, Philosophia Antiqua, 22 (Leiden, 1972).

GRANT, R., *Eusebius as Church Historian* (Oxford, 1980).

——*Miracle and Natural Law in Graeco-Roman and Early Christian Thought* (Amsterdam, 1952).

——'The *Stromateis* of Origen', in J. Fontaine and C. Kannengiesser (eds.), *Epektasis. Mélanges patristiques offerts au Cardinal Jean Daniélou* (Beauchesne, 1972), 285–92.

——'Theological Education at Alexandria', in B. Pearson and J. Goehring (eds.), *The Roots of Egyptian Christianity*, Studies in Antiquity and Christianity (Philadelphia, 1986), 178–89.

GREER, R., *The Captain of our Salvation. A Study in the Patristic Exegesis of Hebrews* (Tübingen, 1973).

GREGG, R. C. (ed.), *Arianism: Historical and Theological Reassessments*, Patristic Monograph Series, 11 (Philadelphia, 1985).

——and GROH, D., 'The Centrality of Soteriology in Early Arianism', *Anglican Theological Review*, 59 (1977), 260–78.

————*Early Arianism. A View of Salvation* (Philadelphia, 1981).

GRILLMEIER, A., *Christ in the Christian Tradition*, Vol. 1: *From the Apostolic Age to Chalcedon (451)*, 2nd rev. edn., tr. J. Bowden (London, 1975).

——'"*Christus licet nobis invitis deus.*" Ein Beitrag zur Diskussion über die Hellenisierung der christlichen Botschaft', in A. M. Ritter (ed.), *Kerygma und Logos. Festschrift für Carl Andresen* (Göttingen, 1979), 226–57.

GRUBER, G., *ΖΩΗ: Wesen, Stufen und Mitteilung des wahren Lebens bei Origenes*, Munich Theological Studies, 2. 23 (Munich, 1962).

GUILLAUMONT, A., 'Évagre et les anathématismes antiorigénistes de 553', *SP* 3 (1961), 213–26.

GULLEY, N., 'The Interpretation of "No one does wrong willingly" in Plato's Dialogues', *Phronesis*, 10 (1965), 82–96.

HADOT, I., 'The Spiritual Guide', in A. H. Armstrong (ed.), *Classical Mediterranean Spirituality* (New York, 1986), 436–59.

HADOT, P., 'La Metáphysique de Porphyre', in *Porphyre*, Entretiens sur l'antiquité classique, 12 (Geneva, 1965), 125–57.

——*Porphyre et Victorinus* (Paris, 1968).

HAHN, F., *The Titles of Jesus in Christology* (London, 1969).

HÄLLSTRÖM, G. af, *Fides simpliciorum according to Origen of Alexandria* (Helsinki, 1984).

HANSON, R. P. C., *The Search for the Christian Doctrine of God* (Edinburgh, 1988).

HARL, M., *Origène et la fonction révélatrice du Verbe incarné*, Patristica Sorbonensia, 2 (Paris, 1958).

——'Recherches sur l'origénisme d'Origène: la 'satiété (κορος) de la contemplation comme motif de la chute des âmes', *SP* 8 (1966), 373–405.

——'Structure et cohérence du Peri Archon', in H. Crouzel, G. Lomiento, and J. Rius-Camps (eds.), *Origeniana*, Quaderni di 'Vetera Christianorum', 12 (Bari, 1975), 11–32.

HARVEY, A. E., *Jesus and the Constraints of History* (London, 1982).

HENGEL, M., *Judaism and Hellenism*, tr. J. Bowden (London, 1974; repr. 1981).

HERON, A., 'Logos, Image, Son: Some Models and Paradigms in Early Christology', in R. W. A. McKinney (ed.), *Creation, Christ, and Culture. Studies in Honour of T. F. Torrance* (Edinburgh, 1976), 44–62.

HOLDCROFT, I. T., 'The Parables of the Pounds and Origen's Doctrine of Grace', *JTS*, NS 24 (1983), 503–4.

HOLZ, H., 'Über den Begriff des Willens und der Freiheit bei Origenes', *Neue Zeitschrift für systematische Theologie*, 12 (1970), 63–84.

HOOD, R. E., *Must God Remain Greek? Afro Cultures and God Talk* (Philadelphia, 1990).

JACKSON, B. DARRELL, 'Sources of Origen's Doctrine of Freedom', *CH* 35 (1966), 13–23.

JUDGE, E. A., and PICKERING, S. R., 'Papyrus Documentation of Church and Community in Egypt to the Mid-Fourth Century', *Jahrbuch für Antike und Christentum*, 20 (1977), 47–71.

KANNENGIESSER, C., *Athanase d'Alexandrie. Evêque et écrivain* (Paris, 1983).

——'The Athanasian Decade 1974–1984: A Bibliographical Report', *TS* 46 (1985), 524–41.

——'Athanasius of Alexandria and the Foundation of Traditional Christology', *TS* 34 (1973), 103–13.

——'Athanasius of Alexandria vs. Arius: The Alexandrian Crisis', in B. Pearson and J. E. Goehring (eds.), *The Roots of Egyptian Christianity*, Studies in Antiquity and Christianity (Philadelphia, 1986), 204–15.

——*Holy Scripture and Hellenistic Hermeneutics in Alexandrian Christology: The Arian Crisis*, ed. I. Lawrence, CHS 41 (Berkeley, Calif., 1982).

——'λογος et νους chez Athanase d'Alexandrie', *SP* 11 (1967), 199–202.

——(ed.), *Politique et théologie chez Athanase d'Alexandrie*, Théologie historique, 27 (Paris, 1974).

KEE, H. G., *Miracle in the Early Christian World* (New Haven, Conn., 1983).

KELLY, J. N. D., *Early Christian Doctrines* (5th edn., Oxford, 1980).

KETTLER, F. H., *Der ursprüngliche Sinn der Dogmatik des Origenes*, Beiheft zur Zeitschrift für die neutestamentliche Wissenschaft und die Kunde der Älteren Kirche, 31 (Berlin, 1966).

——'Die Ewigkeit der geistigen Schopfung nach Origenes', in M. Greschat and J. F. G. Goeters (eds.), *Reformation und Humanismus* (Witten, 1969), 272–97.

KINNEAVY, J. L., *Greek Rhetorical Origins of Christian Faith* (Oxford, 1987).

KOCH, H., *Pronoia und Paideusis: Studien über Origenes und sein Verhältnis zum Platonismus* (Leipzig, 1932).

KOPECEK, T. A., 'Neo-Arian Religion: The Evidence of the *Apostolic Constitutions*', in R. C. Gregg (ed.), *Arianism: Historical and Theological Reassessments*, Patristic Monograph Series, 11 (Philadelphia, 1985), 153–79.

KÜBEL, P., *Schuld und Schicksal bei Origenes, Gnostikern und Platonikern* (Stuttgart, 1973).

LAEUCHI, S., 'Origen's Interpretation of Judas Iscariot', *CH* 22 (1953), 253–65.

LAPIDGE, M., 'Stoic Cosmology', in J. M. Rist (ed.), *The Stoics* (Berkeley, Calif., 1978), 161–85.

LE BOULLUEC, A., 'Controverses au sujet de la doctrine d'Origène sur l'âme du Christ', in L. Lies (ed.), *Origeniana Quarta*, Innsbrucker theologische Studien, 19 (Innsbruck and Vienna, 1987), 223–37.

——*La Notion d'hérésie dans la littérature grecque IIe–IIIe siècles* (2 vols., Paris, 1985).

LEBRETON, J., 'Le Désaccord de la foi populaire et de la théologie savante dans l'Église chrétienne du IIIe siècle', *Revue d'histoire ecclesiastique*, 19 (1923), 481–506; 20 (1924), 5–37.

LIENHARD, J., 'The "Arian" Controversy: Some Categories Reconsidered', *TS* 48 (1987), 415–31.

LIEU, S., *Manichaeism in the Later Roman Empire and Medieval China* (Manchester, 1985).

LLOYD, A. L., 'Emotion and Decision in Stoic Psychology', in J. M. Rist (ed.), *The Stoics* (Berkeley, Calif., 1978), 223–46.

LOENEN, J. H., 'Albinus' Metaphysics: An Attempt at Rehabilitation', *Mnemosyne*, 9 (1956), 296–319.

LORENZ, R., *Arius judaizans? Untersuchungen zur dogmengeschichtlichen Einordnung des Arius* (Göttingen, 1979).

——'Die Christusseele im Arianischen Streit', *Zeitschrift für Kirchengeschichte*, 94 (1983), 1–51.

LOUTH, A., 'Athanasius' Understanding of the Humanity of Christ', *SP* 16.2 (1985), 309–18.

——'The Concept of the Soul in *Contra Gentes — De Incarnatione*', *SP* 13 (1975), 227–31.

——*The Origins of the Christian Mystical Tradition from Plato to Denys* (Oxford, 1981).

——'St. Athanasius and the Greek *Life of Antony*', *JTS*, NS 39 (1988), 504–9.

LOWRY, C. W., 'Did Origen Style the Son a *ΚΤΙΣΜΑ*?', *JTS* 39 (1938), 39–42.

LUHRMANN, D., *Glaube im frühen Christem* (Gutersloh, 1976).

LUIBHEID, C., *Eusebius of Caesarea and the Arian Crisis* (Dublin, 1978).

LYMAN, J. R., 'Arians and Manichees on Christ', *JTS*, NS 40 (1989), 493–503.

——'*Lex Orandi*: Heresy, Orthodoxy, and Popular Religion', in D. Pailin and S. Coakley (eds.), *The Making and Re-Making of Christian Doctrine. Essays in Honour of Maurice Wiles* (Oxford, forthcoming).

——'Substance Language in Origen and Eusebius', in R. C. Gregg (ed.), *Arianism: Historical and Theological Reassessments*, Patristic Monograph Series, 11 (Philadelphia, 1985), 257–66.

MCFAGUE, S., *Metaphorical Theology. Models of God in Religious Language* (Philadelphia, 1982).

MCLEOD, C. W., 'Allegory and Mysticism in Origen and Gregory of Nyssa', *JTS*, NS 22 (1971), 362–79.

MACMULLEN, R., *Christianizing the Roman Empire AD 100–400* (New Haven, Conn., 1984).

MARCUS, W., *Der Subordinationismus als historiologisches Phänomen* (Munich, 1963).

MARKUS, R. A., 'The Problem of Self-Definition: From Sect to Church', in E. P. Sanders (ed.), *Jewish and Christian Self-Definition*, vol. 1 (London, 1980), 1–15.

MEEKS, W., *The First Urban Christians* (New Haven, Conn., 1983).

——*The Moral World of the First Christians* (Philadelphia, 1986).

MEIJERING, E. P., 'Athanasius on the Father as the Origin of the Son', in idem, *God Being History. Studies in Patristic Philosophy* (Amsterdam, 1975), 89–102.

——'Die "Physische Erlösung" in der Theologie des Irenaeus', in idem, *God Being History. Studies in Patristic Philosophy* (Amsterdam, 1975), 39–59.

——'*ΗΝ ΠΟΤΕ ΟΤΕ ΟΥΚ ΗΝ Ο ΥΙΟΣ*: A Discussion on Time and Eternity', in idem, *God Being History. Studies in Patristic Philosophy* (Amsterdam, 1975), 81–8.

——*God Being History. Studies in Patristic Philosophy* (Amsterdam, 1975).

MEIJERING, E. P., 'Irenaeus' Relation to Philosophy in the Light of his Concept of Free Will', in idem, *God Being History. Studies in Patristic Philosophy* (Amsterdam, 1975), 19–30.

——*Orthodoxy and Platonism in Athanasius. Synthesis or Antithesis?* (Leiden, 1968).

——'Wie platonisierten Christen? Zur Grenzziehung zwischen Platonismus, kirchlichen Credo und patrischer Theologie', *VC* 28 (1974), 15–28.

——'Zehn Jahre Forschung zur Thema Platonismus und Kirchenväter', *Theologische Rundschau*, NS 36 (1971), 303–20.

MEREDITH, A., 'Porphyry and Julian Against the Christians', *ANRW* II. 23. 2 (1980), 1119–49.

MOORE, G. F., 'Fate and Free Will in the Jewish Philosophies according to Josephus', *HTR* 22 (1929), 371–89.

MOULE, C. F. D., 'The Borderlands of Ontology in the New Testament', in B. Hebblethwaite and S. Sutherland (eds.), *The Philosophical Frontiers of Christian Theology* (London, 1982), 1–11.

MOULTON, J. H., *The Vocabulary of the New Testament* (London, 1918–30).

MOUTSOULAS, E. D., 'La Lettre d'Athanase d'Alexandrie à Epictète', in C. Kannengiesser (ed.), *Politique et théologie chez Athanase d'Alexandrie*, Théologie historique, 27 (Paris, 1974), 313–33.

MÜHLENBERG, E., 'The Divinity of Jesus in Early Christian Faith', *SP* 17 (1982), 136–46.

——'Vérité et bonté de Dieu. Une interpretation de *De Incarnatione*, chapitre vi, en perspective historique', in C. Kannengiesser (ed.), *Politique et théologie chez Athanase d'Alexandrie*, Théologie historique, 27 (Paris, 1974), 215–30.

NAUTIN, P., '"Je suis celui qui est" (Exode 3. 14) dans la théologie d'Origène', in idem, *Dieu et l'être. Exégèses d'Exode 3. 14 et de Coran 20. 11–24*, Études Augustiniennes (Paris, 1978), 109–19.

——*Origène. Sa vie et son ouevre*, Christianisme antique, 1 (Paris, 1977).

NEWTON, I., *Theological Manuscripts*, ed. H. McLachlan (Liverpool, 1950).

NOCK, A. D., 'Studies in the Graeco-Roman Beliefs of the Empire', in Z. Stewart (ed.), *Arthur Darby Nock. Essays on Religion and the Ancient World*, vol. 1 (Cambridge, Mass., 1972), 33–48.

Norris, R., *God and the World in Early Christian Thought* (New York, 1965).

——*Manhood and Christ. A Study in the Christology of Theodore of Mopsuestia* (Oxford, 1963).

——'The Problem of Human Identity in Patristic Christological Speculation', *SP* 17.1 (1982), 147–59.

——'The Transcendence and Freedom of God: Irenaeus, the Greek Tradition and Gnosticism', in W. R. Schoedel and R. C. Wilken (eds.), *Early Christian Literature and the Classical Intellectual Tradition. In honorem Robert M. Grant*, Théologie historique, 54 (Paris, 1979), 87–100.

O'Cleirigh, P. M., 'The Meaning of Dogma in Origen', in E. P. Sanders (ed.), *Jewish and Christian Self-Definition*, vol. 1 (London, 1980), 201–16.

Opitz, H. G., 'Euseb von Caesarea als Theologe: Ein Vortrag', *ZNW* 34 (1935), 1–19.

Orbe, A., *Hacia la primera teologia de la procesion del Verbo. Estudios valentinianos*, vol. 1, Analecta Gregoriana, 99–100 (Rome, 1958).

Otis, B., 'Cappodocian Thought as a Coherent System', *Dumbarton Oaks Papers*, 12 (1958), 95–124.

Pack, R., 'A Passage in Alexander of Aphrodisias Relating to the Theory of Tragedy', *American Journal of Philology*, 58 (1937), 418–36.

Pannenberg, W., 'The Appropriation of the Philosophical Concept of God as a Dogmatic Problem of Early Christian Theology', in idem, *Basic Questions in Theology*, vol. 2, tr. J. Baker (London, 1971), 119–83.

Patterson, L., 'Methodius, Origen, and the Arian Dispute', *SP* 17.2 (1982), 912–23.

Pearson, B. A., and Goehring, J. E. (eds.), *The Roots of Egyptian Christianity*, Studies in Antiquity and Christianity (Philadelphia, 1986).

Pépin, J., *Idées grecques sur l'homme et sur Dieu*, Collection d'études anciennes de l'Association Guillaume Budé (Paris, 1971).

——*Théologie cosmique et théologie chrétienne* (Paris, 1964).

Perkins, P., 'Ordering the Cosmos: Irenaeus and the Gnostics', in C. W. Hedrick and R. Hodgson (eds.), *Nag Hammadi, Gnosticism, and Early Christianity* (Peabody, Mass., 1986), 221–38.

Petterson, A. L., 'The Questioning Jesus in Athanasius' *Contra Arianos* III', in R. C. Gregg (ed.), *Arianism: Historical and Theological Reassessments*, Patristic Monograph Series, II (Philadelphia, 1985), 243–56.

Places, E. des, *Eusèbe de Césarée Commentateur. Platonisme et écriture sainte* (Paris, 1982).

Prestige, G. L., ''Ἀγεν[ν]ητος and γεν[ν]ητος and Kindred Words in Eusebius and the Early Arians', *JTS*, 24 (1923), 486–96.

——*God in Patristic Thought* (2nd edn., London, 1952).

Price, S. R. F., *Rituals and Power. The Roman Imperial Cult in Asia Minor* (Cambridge, 1984).

Refoule, F., 'La Christologie d'Évagre et l'origénisme', *Orientalia Christiana Periodica*, 27 (1961), 221–66.

RICHARD, M., 'Saint Athanase et la psychologie du Christ selon les Ariens', *Mélanges de science religieuse*, 4 (1947), 5–54.

RICKEN, F., 'Die Logoslehre des Eusebios von Caesarea und der Mittelplatonismus', *Theologie und Philosophie*, 42 (1967), 341–58.

——'Nikaia als Krisis des altchristlichen Platonismus', *Theologie und Philosophie*, 44 (1969), 321–41.

——'Zur Rezeption der platonischen Ontologie bei Eusebios von Kaisareia, Areios und Athanasios', *Theologie und Philosophie*, 53 (1973), 321–52.

RIEDMATTEN, H., *Les Actes du procès de Paul de Samosate* (Fribourg, 1952).

RIST, J. M., 'Basil's "Neoplatonism": Its Background and Nature', in P. Fedwick (ed.), *Basil of Caesarea: Christian, Humanist, Ascetic. A Sixteen-Hundredth Anniversary Symposium*, vol. 1 (Toronto, 1981), 137–219.

——'Beyond Stoic and Platonist: A Sample of Origen's Treatment of Philosophy (*Contra Celsum* 4. 62–70)', in H. D. Blume and F. Mann (eds.), *Platonismus und Christentum. Festschrift für Heinrich Dörrie*, Jahrbuch für Antike und Christentum, suppl. vol. 10 (1983), 228–38.

——'The Greek and Latin Texts in *De Principiis* Book III', in H. Crouzel (ed.), *Origeniana* (Bari, 1975), 97–111.

——*Plotinus. The Road to Reality* (Cambridge, 1967).

——'Prohairesis: Proclus, Plotinus *et alii*', in *De Jamblique à Proclus*, Entretiens sur l'antiquité classique, 21 (1975), 103–17.

——*Stoic Philosophy* (Cambridge, 1969; repr. 1980).

——(ed.), *The Stoics* (Berkeley, Calif., 1978).

RITTER, A. M., 'Platonismus und Christentum in der Spätantike', *Theologische Rundschau*, NS 49 (1984), 31–56.

RIUS-CAMPS, J., *El Dinamismo trinitario en la divinizacion de los seres racionales segun Origenes*, Orientalia Christiana Analecta, 188 (Rome, 1970).

——'La Suerte final de la naturaleza corporea segun el Peri Archon de Origenes. Formulacion fluctuante entre el dato revelado y los presupuestos filosoficos de un sistema', *SP* 14 (1976), 167–79.

ROLDANUS, J., *Le Christ et l'homme dans la théologie d'Athanase d'Alexandrie. Étude de la conjonction de sa conception de l'homme avec sa christologie*, Studies in the History of Christian Thought, 4 (Leiden, 1968).

RONDEAU, M.-J., 'Une nouvelle preuve de l'influence littéraire d'Eusèbe de Césarée sur Athanase: l'interpretation des psaumes', *Recherches de science religieuse*, 56 (1968), 385–434.

ROSE, E., *Die Manïchaische Christologie* (Wiesbaden, 1979).

RUDOLPH, K., *Gnosis. The Nature and History of Gnosticism*, tr. R. M. Wilson (New York, 1983).

RUETHER, R. RADFORD, *To Change the World. Christology and Cultural Criticism* (New York, 1981).

SAFFREY, H., 'Les Extraits du *περὶ τἀγαθου* de Numenius dans le livre XI de la *Préparation évangelique* d' Eusèbe d' Césarée', *SP* 13 (1975), 46–51.

SAMPLE, R., 'The Christology of the Council of Antioch (268 C.E.) Reconsidered', *CH* 48 (1979), 18–26.

SCHARLEMANN, R. P., 'Theological Models and their Construction', *Journal of Religion*, 53 (1973), 69–70.

SCHREITER, R. J., *Constructing Local Theologies* (Maryknoll, NY, 1986).

SIEBEN, H. J., 'Herméneutique de l'exégèse dogmatique d'Athanase', in C. Kannengiesser (ed.), *Politique et théologie chez Athanase d'Alexandrie*, Théologie historique, 27 (Paris, 1974), 195–214.

SIMONETTI, M., 'Note sulla teologia trinitaria di Origene', *Vetera Christianorum*, 8 (1971), 273–307.

——'Sull'interpretazione di un passo del *De Principiis* di Origene (I. 3. 5–8)', *Rivista di Cultura Classica e Medievale*, 6 (1964), 15–32.

SLUSSER, M., 'Athanasius, *Contra Gentes*, and *De Incarnatione*: Place and Date of Composition', *JTS*, NS 37 (1986), 114–17.

SMITH, A., *Porphyry's Place in the Neoplatonic Tradition* (The Hague, 1974).

SMITH, J. Z., *Drudgery Divine. On the Comparison of Early Christianities and the Religions of Late Antiquity* (Chicago, 1990).

SMITH, M., 'A Hidden Use of Porphyry's *History of Philosophy* in Eusebius' *Praeparatio Evangelica*', *JTS*, NS 39 (1988), 500–4.

SMITH, R. C., and LOUNIBOS, J. (eds.), *Pagan and Christian Anxiety. A Response to E. R. Dodds* (Lanham, Md., 1984).

SORABJI, R., *Time, Creation, and the Continuum* (New York, 1983).

STEAD, G. C., 'Athanasius' Earliest Written Work', *JTS*, NS 39 (1988), 76–91.

——*Divine Substance* (Oxford, 1977).

——'"Eusebius" and the Council of Nicaea', *JTS*, NS 24 (1973), 85–100.

——'The Freedom of the Will and the Arian Controversy', in H. D. Blume and F. Mann (eds.), *Platonismus und Christentum. Festschrift für Heinrich Dörrie*, Jahrbuch für Antike und Christentum, suppl. vol. 10 (1983), 245–57.

——'Individual Personality in Origen and the Cappadocian Fathers', in U. Bianchi and H. Crouzel (eds.), *Arché e Telos. L'antropologia di Origene di Gregorio di Nissa. Analisi storico-religiosa*, Studia Patristica Mediolanensia, 12 (Milan, 1981), 170–87.

STEAD, G. C., 'The Platonism of Arius', *JTS*, NS 15 (1964), 16–31.
——'Rhetorical Method in Athanasius', *VC* 30 (1976), 121–37.
——'The Scriptures and the Soul of Christ in Athanasius', *VC* 36 (1982), 233–50.
STIEGMANN, E., 'Rabbinic Anthropology', *ANRW* II. 19. 2 (1979), 487–579.
STOUGH, C., 'Stoic Determinism', in J. M. Rist (ed.), *The Stoics* (Berkeley, Calif., 1978), 203–31.
TARAN, L., 'The Creation Myth in Plato's *Timaeus*', in J. P. Anton and G. L. Kustas (eds.), *Essays in Ancient Greek Philosophy* (Albany, NY, 1971), 372–407.
TELFER, W., 'ΑΥΤΕΧΟΥΣΙΑ', *JTS*, NS 3 (1957), 123–9.
TETZ, M., 'Zur Biographie des Athanasius von Alexandrien', *Zeitschrift für Kirchengeschichte*, 90 (1979), 58–92
TODD, R., 'Monism and Immanence: The Foundations of Stoic Physics', in J. M. Rist (ed.), *The Stoics* (Berkeley, Calif., 1978), 137–60.
TORJESEN, K. J., *Hermeneutical Procedure and Theological Method in Origen's Exegesis*, Patristic Texts and Studies, 28 (Berlin, 1985).
——'Origen's Interpretation of the Psalms', *SP* 17.2 (1982), 944–58.
——'The Teaching Function of the Logos: Athanasius, *De Incarnatione* xx–xxxii', in R. C. Gregg (ed.), *Arianism: Historical and Theological Reassessments*, Patristic Monograph Series, II (Philadelphia, 1985), 213–26.
TRACY, D., and LASH, N. (eds.), *Cosmology and Theology*, Concilium, 166 (New York, 1983).
TRAKATELLIS, D. C., *The Pre-Existence of Christ in the Writings of Justin Martyr*, Harvard Dissertations in Religion, 6 (Missoula, Mont., 1976).
TRIGG, J. W., 'The Charismatic Intellectual: Origen's Understanding of Religious Leadership', *CH* 50 (1981), 5–19.
——*Origen. The Bible and Philosophy in the Third Century Church* (Atlanta, 1983).
ULANSEY, D., *The Origins of the Mithraic Mysteries. Cosmology and Salvation in the Ancient World* (Oxford, 1989).
VALENTASIS, R., *Spiritual Guides of the Third Century* (Minneapolis, 1991).
VEYNE, P., *Did the Greeks Believe their Myths?*, tr. P. Wissing (Chicago, 1988).
VOELKE, A.-J., *L'Idée de volonté dans le stoicisme* (Paris, 1973).
VOGEL, C. J. DE, 'Platonism and Christianity: A Mere Antagonism or a Profound Common Ground?' *VC* 39 (1985), 1–62.
WALKER, J. B., 'Covenance épistémologique de l'"Homoousion" dans la théologie d'Athanase', in C. Kannengiesser (ed.), *Politique et théologie*

chez Athanase d'Alexandrie, Théologie historique, 27 (Paris, 1974), 255–75.

WALLACE-HADRILL, D. S., *Eusebius of Caesarea* (Westminster, 1961).

WEBER, A., *'APXN. Ein Beitrag zur Christologie des Eusebius von Caesarea* (Rome, 1964).

WHITTAKER, J., '*Ἐπέκεινα νοῦ καὶ οὐσίας*', *VC* 23 (1969), 91–104.

——'The Historical Background of Proclus' Doctrine of the Ἀυθυπόστατοι in *De Jamblique à Proclus*, Entretiens sur l'antiquité classique, 21 (Geneva, 1975), 193–230.

——'Moses Atticizing', *Phoenix* 21 (1967), 196–201.

WILES, M. F., 'Eternal Generation', *JTS*, NS 13 (1961), 284–91.

——'Eunomius: Hair-Splitting Dialectician or Defender of the Accessibility of Salvation?', in R. Williams (ed.), *The Making of Orthodoxy* (Cambridge, 1989), 157–69.

——*The Making of Christian Doctrine* (Cambridge, 1967).

WILKEN, R., 'Wisdom and Philosophy in Early Christianity', in idem (ed.), *Aspects of Wisdom in Judaism and Early Christianity* (Notre Dame, Ind., 1975), 143–68.

WILLIAMS, C., *The Descent of the Dove* (New York, 1939).

WILLIAMS, R., *Arius. Heresy and Tradition* (London, 1987).

——'Does it Make Sense to Speak of a Pre-Nicene Orthodoxy?', in idem (ed.), *The Making of Orthodoxy. Essays in Honour of Henry Chadwick* (Cambridge, 1990), 1–23.

——(ed.), *The Making of Orthodoxy. Essays in Honour of Henry Chadwick* (Cambridge, 1990).

——'Origen on the Soul of Jesus', in R. P. C. Hanson (ed.), *Origeniana Tertia* (Rome, 1985), 131–7.

WINSTON, D., 'The Book of Wisdom's Theory of Cosmogony', *History of Religions*, 2 (1971), 185–202.

——'Creation *Ex Nihilo* Revisited: A Reply to Jonathan Goldstein', *Journal of Jewish Studies*, 37 (1986), 88–91.

——'Freedom and Determinism in Philo of Alexandria', *Studia Philonica*, 3 (1974–5), 47–70.

WISSE, F., 'The Use of Early Christian Literature as Evidence for Inner Diversity and Conflict', in C. W. Hedrick and R. Hodgson (eds.), *Nag Hammadi, Gnosticism, and Early Christianity* (Peabody, Mass., 1986), 177–90.

YOUNG, F., *From Nicaea to Chalcedon* (Philadelphia, 1983).

GENERAL INDEX

Adam and Eve 33, 35, 143
Albinus 13, 107, 131
Alexander of Alexandria 82, 86, 93, 126, 139, 147, 148
Alexander of Lycopolis 15
Alexandria 8, 40–2, 125–6
Amand de Mendieta, E. A. 17 n., 52
anthropology 28–36, 58–69, 100–6, 141–6; body 61–2, 103, 120, 143, 153; freedom of will 31–6, 59–63, 68, 102, 144; image of God 35, 62, 101–2, 142; intellect 29, 62, 142
Apollinarius 154, 155
apologetics 10, 22, 86–7
Apuleius 16
Arianism 7, 97, 112, 126–7, 129, 133, 147, 149, 155–6
Aristotelians 15, 29
Arius 7, 82, 85, 93, 127; epistemology 129, 131; generation of Son 97, 110, 135, 148
Armstrong, A. H. 7, 38, 160
Asterius 82, 86
asceticism 105, 125–6, 140, 143, 159
Athanasius 8, 113, 121
Atticus 14, 88
Augustine 8, 61, 62
αὐτο-prefixes 73, 95–6, 102, 113, 149

Barnard, L. 25 n., 37
Barnes, T. D. 86
Benz, E. 51
Berchman, R. M. 52
biblical, *see* Scripture
Brown, P. 3, 46 n., 143

Caesarea 41–2, 84–5, 88
Cappadocians 68 n., 150
Celsus 11, 18, 36
changeability 63, 102, 104, 138, 141–2
Christ 20–8, 69–80, 107–22, 146–58; agency of Logos 25, 37, 72–3, 77, 116–17, 121, 152, 155; exemplar 27, 76–7, 79, 157, 161; generation from Father 70–2, 108–14, 148; Gethsemane 19, 79, 119, 121, 157; high priest 74–6, 116; human soul 74–7, 85, 117; obedience of Jesus 19, 27, 57, 76–7, 115–16, 119–21, 156–7; πρόσωπον 119–22, 154–7; second Adam 27, 35; *see also* hierarchy, Incarnation
Clement 135
Constantine 82–3, 86, 98
Constantius 127
Coptic Christianity 126
cosmology 6, 15–17, 25, 27; as a theological model 5, 12; mediation as 'Platonic' 24, 37, 83, 107–8
creation 22–3, 26, 54–5, 94–7, 137–8; *ex nihilo* 5, 18, 23, 37, 55, 96, 137
Crouzel, H. 74, 76 n.

Dihle, A. 5 n., 18 n., 32
Diocletian 83, 84
divinization, *see* soteriology
Dörrie, H. 1, 17, 37
Douglas, M. 4 n., 162
Drake, H. 86, 98 n.

emanation 14, 70
Epictetus 16
episcopacy 41 n., 125–6
Eusebius 8, 126, 128, 135–6, 144, 148, 153, 156
exegesis, *see* Scripture

faith 19, 20, 22, 31–2, 34, 36, 129, 142
Fall 30–1, 33, 35, 60–1, 100–7, 138–9, 141–2
fatalism 17, 29, 161
Father, *see* God

Galen 13, 18
Gnosticism 4, 10, 134; anthropology 31, 34, 44, 59, 62, 79; creation 26, 48, 50, 137
God: as ἀγένητος 23, 47, 93–4, 96–7, 112, 134, 150; as creator 21–3, 26, 54, 91, 93–5, 135, 137–8; as Father 23, 27, 69, 111–13, 133–4; foreknowledge 57, 76, 101; will in history

God (*cont.*):
20–1, 35, 52–3, 92, 99, 104, 139; will and nature 52–3, 55–6, 95, 135–6, 151; *see also* hierarchy, transcendence
goodness 50, 76, 91, 95, 135
Graeco-Roman religion 10, 16, 31
grace 21, 49, 66, 100
Gregg, R. and D. Groh 7, 125, 127, 139, 143, 156
Gregory Thaumaturgus 42
Grillmeier, A. 75
Gruber, G. 72

Harl, M. 61, 72
Hellenization of Christianity 6, 85, 162
heresy 2, 8, 43, 127
hierarchy of Father and Son 6–7, 24, 28, 37, 43, 72–3, 111, 115–16, 149
history 19, 35, 84, 88, 99, 104, 128, 131, 139–40
Holtz, H. 51–2

Incarnation 4, 24, 27, 73, 117, 153
Irenaeus 25–7, 34–6, 61, 96, 104, 134, 137
Isis 16, 31

Jesus, *see* Christ
Judaism 18–20, 22, 31–2, 42, 45, 47
Judas 57, 66–7
Justin 22–5, 33–4

Kannengiesser, C. 125, 127, 141, 146 n.
Kettler, F. 51
Koch, H. 52
knowledge 20, 22, 34, 37, 64, 66, 104, 129–31

Lienhard, J. 7, 160
Logos, *see* Christ
Lorenz, R. 7

McFague, S. 4
Manichees 15, 97, 110, 148; anthropology 102, 146; epistemology 129, 131
Marcellus of Ancyra 82, 86, 109, 115, 155
martyrs 21, 79, 84, 157
mediation, *see* cosmology
Meeks, W. 3, 162 n.
Meijering, E. P. 2 n., 125, 128 n.
Methodius 97, 101–2, 110, 137, 139, 142
model 4

nature: divine 15, 51, 69, 150; in contrast to will 7, 9, 95; ὁμοούσιος 48, 59, 69, 97, 150; οὐσία 51, 69, 150; πνεῦμα 15, 51, 65; φύσις 15, 109, 150
Nestorius 7
Nicaea 6, 86, 97, 127, 150
Numenius 13, 14, 49, 88, 96, 98–9, 107, 111

obedience 32–3, 66; *see also* Christ, will
Origen 85, 88–9, 100–1, 111, 113–14, 116, 136, 142, 150–2, 156–7

paganism, *see* Graeco-Roman religion
Pamphilus 82, 84–5, 88
participation (μετέχω) 48, 64, 114, 134
Paul 60–1
Pelagius 8
Philo 18, 48–9, 107
philosophy 2, 8, 10, 16–17, 41–2; *see also* Aristotelians, Platonism, Stoicism
piety 2, 8, 105
Platonism 12–15, 29, 30, 56, 88–9, 131, 135, 142, 146, 159; *Timaeus* 13–14, 23, 54, 95
Plotinus 14, 17, 30, 89, 95, 99, 110, 131, 136
Porphyry 14, 17, 30–1, 42, 85, 87, 89, 96, 136
Plutarch 16
power, divine 14–15, 16–17, 31, 56, 79, 87–8, 92, 99, 105, 118, 138–9; *see also* transcendence
pre-existence of souls 44, 60, 101
progress 67–8, 76, 145
providence (πρόνοια) 17, 30, 98–9, 139; *see also* God
revelation 20–1, 38, 129–31, 158
Richard, M. 155
Ricken, F. 2, 98, 99, 107–8, 127
Rist, J. M. 43
Rius Camps, J. 51
Roldanus, J. 139, 141, 143

Scripture: contrasted to philosophy 2, 12, 17, 94; cosmology 2, 17–20; exegesis 18, 129, 153–4; synthesis with philosophy 49, 108–9, 117, 129, 133; Exodus 3: 14 49, 93, 95, 133; Mark 10: 18 43, 49, 50, 73; Psalm 45: 7–8 76, 156; Wisdom 7: 26–8 50, 70, 73

Seneca 16
Simonetti, M. 51–2
Son, *see* Christ
soteriology: and cosmology 6–8, 25, 27, 122; divinization 67, 82, 105–6, 122, 144–5, 158, 161; *see also* will
Stead, G. C. 36 n., 66 n., 103, 125, 133, 136, 145, 149, 150
Stoicism 14–16, 18, 29, 32, 45, 56, 62, 64, 72
subordinationism, *see* hierarchy
substance, *see* nature

teacher 22, 40–41, 45, 81, 128
Theophilus of Antioch 32
transcendence 13–15, 17, 47, 87, 131; in Christian theology 21, 27, 47, 87, 91, 93, 98–9, 131
Trinity 50–2
Trakatellis, D. C. 24

Veyne, P. 124
Vogel, C. J. De 127

Wilken, R. 10
will 13–14, 19–20, 29–30; *αὐτεξούσιος* 30, 32–3, 35–6, 63, 100, 102, 104, 138, 146, 155–6; *βουλή*, *βούλησις* *βούλομαι* 13–14, 19, 23–4, 53, 79, 96, 110, 136, 151; *θέλημα*, *θέλησις*, θέλω 14, 19, 23–4, 53, 78–9, 94–5, 136, 138, 151; *προαίρεσις* 29–30, 33, 63, 74–5, 97, 105, 110, 144; as mere moralism 8, 59, 100, 161; in divinization 63, 65–7, 78, 106, 139, 144; knowledge 29, 33, 61, 64–7; *see also* anthropology, Christ, God
Williams, R. 8, 127, 161
women 40, 57, 91, 105, 125
world-view 3, 4

The manufacturer's authorised representative in the EU for product safety is Oxford University Press España S.A. of El Parque Empresarial San Fernando de Henares, Avenida de Castilla, 2 – 28830 Madrid (www.oup.es/en or product.safety@oup.com). OUP España S.A. also acts as importer into Spain of products made by the manufacturer.

Printed in the USA/Agawam, MA
April 30, 2026

904411.068